Plant Genomes

Genome Dynamics

Vol. 4

Plant Genomes

Volume Editor

Jean-Nicolas Volff *Lyon*

12 figures, 2 in color, and 2 tables, 2008

KARGER

Basel · Freiburg · Paris · London · New York · Bangalore · Bangkok · Shanghai · Singapore · Tokyo · Sydney

Prof. Jean-Nicolas Volff
Institut de Génomique Fonctionnelle de Lyon
Ecole Normale Supérieure de Lyon
46 allée d'Italie
F-69364 Lyon Cedex 07 (France)

Library of Congress Cataloging-in-Publication Data

Plant genomes / volume editor, Jean-Nicolas Volff.
p. ; cm. – (Genome dynamics, ISSN 1660–9263 ; v. 4)
Includes bibliographical references and indexes.
ISBN 978–3–8055–8491–3 (hard cover : alk. paper)
1. Plant genomes. I. Volff, Jean-Nicolas. II. Series.
[DNLM: 1. Genome, Plant. 2. Evolution, Molecular. 3. Variation (Genetics) QK 981 P713 2008]
QK981.P535 2008
572.8′62–dc22 2008007967

Bibliographic Indices. This publication is listed in bibliographic services, including Current Contents® and Index Medicus.

Disclaimer. The statements, options and data contained in this publication are solely those of the individual authors and contributors and not of the publisher and the editor(s). The appearance of advertisements in the book is not a warranty, endorsement, or approval of the products or services advertised or of their effectiveness, quality or safety. The publisher and the editor(s) disclaim responsibility for any injury to persons or property resulting from any ideas, methods, instructions or products referred to in the content or advertisements.

Drug Dosage. The authors and the publisher have exerted every effort to ensure that drug selection and dosage set forth in this text are in accord with current recommendations and practice at the time of publication. However, in view of ongoing research, changes in government regulations, and the constant flow of information relating to drug therapy and drug reactions, the reader is urged to check the package insert for each drug for any change in indications and dosage and for added warnings and precautions. This is particularly important when the recommended agent is a new and/or infrequently employed drug.

www.karger.com
Printed in Switzerland on acid-free and non-aging paper (ISO 9706) by Reinhardt Druck, Basel
ISSN 1660–9263
ISBN 978–3–8055–8491–3

Contents

Preface

The fourth volume of 'Genome Dynamics' is dedicated to 'Plant Genomes'. In 2000, the sequencing of the genome of the thale cress *Arabidopsis thaliana* has opened a new era of plant genomics. In the mean time, genome sequences have been completed for additional plants including rice and grapevine, and an important amount of data is already available for many other species. As a consequence, comparative genomics has completely modified our vision of the structure and evolution of plant genomes, with important implications for plant physiology, developmental biology, genetics and evolution as well as obvious applications in the field of agriculture. This volume of 'Genome Dynamics' aims to reflect the astonishing plasticity of plant genomes and to present the molecular and evolutionary processes involved in genome and biological diversity. Particular emphasis is given on the evolutionary impact of genome duplications and transposable elements. Recent progress in the understanding of the evolution of protein-coding gene families and microRNAs are reviewed, with specific focus on the evolutionary consequences of gene duplication. Mechanisms involved in size variations are also discussed, as well as new insights into the structure and evolution of sex chromosomes and centromeres. Analyses included in this book range from wide comparisons between very divergent plant species to deep studies of specific models such as rice and maize.

All papers published in 'Genome Dynamics' are reviewed according to classical standards. I would like to thank all contributors and referees involved in this book, Michael Schmid and his team, as well as Karger Publishers for their invaluable help during the preparation of this volume.

Jean-Nicolas Volff
Lyon, December 2007

Volff J-N (ed): Plant Genomes.
Genome Dyn. Basel, Karger, 2008, vol 4, pp 1–12

Paleopolyploidy and its Impact on the Structure and Function of Modern Plant Genomes

A.H. Paterson

Plant Genome Mapping Laboratory, University of Georgia, Athens, Ga., USA

Abstract

Partial or complete genome duplication is a punctuational event in the evolutionary history of a lineage, with permanent consequences for all descendants. Careful analysis of burgeoning cDNA and genomic sequence data have underlined the importance of genome duplication in the evolution of biological diversity. Of singular importance among the consequences of paleopolyploidy is the extensive loss (or degradation beyond recognition) of duplicated genes. Gene loss complicates genome comparisons by fragmenting ancestral gene orders across multiple chromosomes, and may also link genome duplication to speciation. The recent discovery in angiosperms of gene functional groups that are 'duplication-resistant', i.e. which are preferentially returned to singleton status following genome duplications, adds a new dimension to classical views that focus on the potential advantages of genome duplication as a source of genes with new functions. The surprisingly conservative evolution of coding sequences that are preserved in duplicate, suggests still additional new dimensions in the spectrum of fates of duplicated genes. Looking forward, their many independent genome duplications, together with extensive sets of computational and experimental tools and resources, suggest that the angiosperms may play a major role in clarifying the structural, functional and evolutionary consequences of paleopolyploidy.

Prevalence and Phylogenetic Distribution of Paleopolyploidy

Partial or complete genome duplication is a punctuational event in the evolutionary history of a lineage, with permanent consequences for all descendants – if the lineage survives. Most higher organisms pass through different ploidy levels at different stages of development [1, 2] and are presumed to continuously produce aberrant unreduced gametes at low rates. However, the extreme rarity of genome duplication in the evolutionary history of extant lineages, occurring only once in many (sometimes hundreds of) millions of years, shows that the vast majority of genome duplication events quickly go extinct.

In contrast to most animals and microbes in which genome duplication is so ancient as to leave only tenuous signal and low levels of gene duplication, all angiosperms are thought to have been affected by an ancient duplication event [3] and the vast majority have been affected by additional, more recent events (fig. 1). Building on long-standing cytological evidence and hints based on isozymes [4, 5], and DNA markers [6–10], evidence of segmental or whole-genome duplications derives from non-random patterns in the arrangement of duplicated genes in genomic sequences. Each of the three angiosperm genomes fully-sequenced to date show evidence of independent whole-genome duplications [3, 11, 12]. Deviations from the expected L-shaped age distributions associated with large EST sets imply additional genome duplications [13–15] – albeit failing to detect some events that have been demonstrated at the whole-genome level [11]. One can only anticipate that the number of ancient duplications known will continue to grow as our sequence coverage of the angiosperms improves, and methods for identifying genome duplications are improving [16].

In rare instances, fossil evidence of paleopolyploidization events has been found [17, 18] – however, dating of most ancient duplications relies on molecular evidence. A first approximation of the age of a genome duplication can be obtained using a 'molecular clock' such as the neutral divergence rate associated with one or more well-studied genes [19]. The use of such a rate permits one to estimate the duration of reproductive isolation that would be necessary to explain the average degree of DNA or protein sequence divergence between pairs of genes. Well-known limitations of molecular clocks [20] can cause striking incongruities with fossil-based dates in some cases [12]. The recent demonstration that ancient duplicated genes appear to undergo some degree of concerted evolution in yeast [21] and perhaps also in angiosperms [22] implies that clock-based approaches may chronically under-estimate the ages of genome duplications.

A dating method that mitigates some weaknesses of clock-based approaches (such as varying evolutionary rates of different genes) is to evaluate large numbers of gene trees including a pair of duplicated genes, a taxon of interest, and an outgroup, to determine whether duplication of the pair predated or postdated their divergence from the taxon of interest [23]. Duplications can be plotted phylogenetically by analyzing large numbers of gene trees and focusing on frequencies of 'internal trees' (in which a gene of interest is more closely related to one member of a duplicate pair than is the other member of the pair, thus which are less prone than 'external' trees to artifacts such as failure to identify the true homolog). For example, a current understanding of the phylogeny of

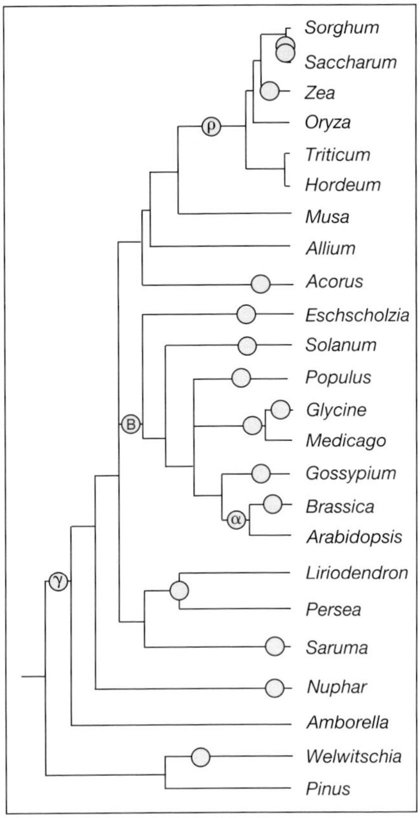

Fig. 1. Phylogeny of known or suspected ancient duplications in the angiosperms, revised from [50], to incorporate additional data from several sources [14, 15, 75, 76]. Lengths of branches and exact locations of suspected genome duplications on branches are approximate, attempting to accurately portray current thinking about relative ages but not following a specific single scale. Greek letters indicate genome duplications identified based on whole-genome analysis while circles indicate additional duplications based on EST data.

angiosperm genome duplications is shown (fig. 1). This is especially useful in formulating meaningful comparisons of duplicated to non-duplicated genomes [24], as shown (fig. 2). For example, the sorghum genome would serve as an excellent 'outgroup' for analysis of the consequences of independent duplications in maize and sugarcane, respectively, that each occurred since the divergence of these lineages from sorghum [25, 26].

Consequences of Paleopolyploidy for Genome Organization

Polyploid formation is a punctuational event in the evolutionary history of a lineage, with permanent and sweeping consequences for all descendant lineages. The ability to 'synthesize' newly-polyploid plants by artificial crosses and chromosomal manipulation using colchicine, has revealed striking immediate

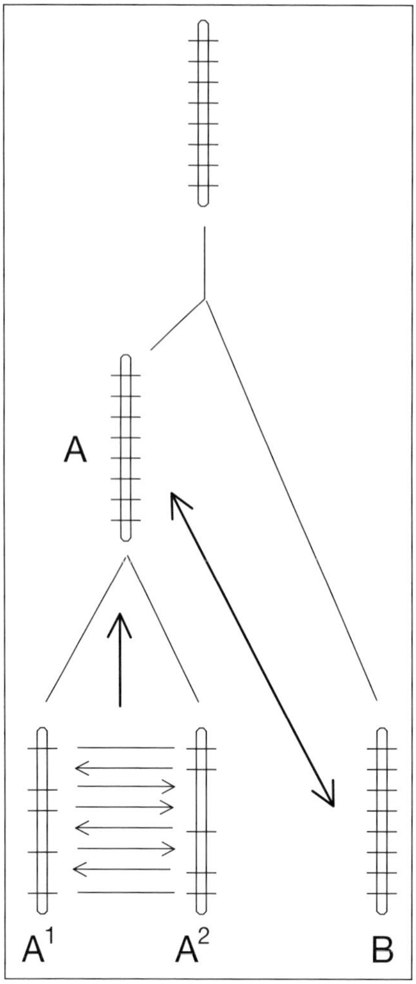

Fig. 2. Understanding the relationship between genome duplication and taxon divergence is key to meaningful genome comparisons. Genome duplication and associated gene loss in a lineage leads to fractionation of ancestral gene sets across multiple chromosomes (A^1, A^2). Comparison of this genome to other related genomes is made more meaningful by using the modern gene orders (as shown by horizontal arrows) to infer the probable gene content and order of a hypothetical ancestral genome (chromosome: A). This inferred hypothetical ancestor often provides for more meaningful comparisons to other genomes. While the figure illustrates a 1:2 comparison, in which genome duplication has occurred in only one lineage since the divergence of the two lineages, one can envision scenarios in which each lineage has duplicated, thus requiring 2:2 comparison.

reactions of genomes to duplication. These reactions include loss and restructuring of low-copy DNA sequences [27–32], activation of genes and retrotransposons [33, 34], gene silencing [35–38] and subfunctionalization of gene expression patterns [39, 40].

There exists virtually no data to distinguish whether immediate reactions of genomes to duplication provide raw material for the beginnings of adaptation to GD, or are symptomatic of imminent extinction. The extinction hypothesis seems more likely, given that unreduced gametes are produced more or less continuously but only a tiny fraction result in successful lineages. For example, dramatic early-generation mutations in synthetic *B. napus* [41] are not paralleled in naturally-occurring forms [42].

Of singular importance among the consequences of paleopolyploidy is the extensive loss (or degradation so as to be no longer recognizable) of duplicated genes. Indeed, in mammals thought to be paleopolyploid, gene loss is so extensive that duplication can only be resolved over 1–5% of the genomes [43–47]. In *Tetraodon* and other ray-finned fish lineages that trace to a common paleopolyploid ancestor that existed perhaps 350 million years ago [48, 49], duplicated copies are still present for about 20% of genes. The more recent duplications in angiosperms within the past 60–70 million years or less are associated with ~30% retention of duplicated gene copies [3, 50].

Although numerous data suggest that post-polyploidization gene loss is far from random (see below), even random gene loss may link genome duplication to speciation. Gene loss resulting in microchromosomal changes would segregate as null homozygotes (missing genes) in F_1 gametes and F_2 zygotes and may cause reproductive isolation [51, 52], as has been tentatively supported in yeast [53].

Post-polyploidization gene loss together with other mechanisms such as translocation fragments ancestral linkage arrangements, creating networks of synteny and colinearity that are distributed across multiple chromosomes [54], that necessitate modification of traditional 'one-to-one' genome comparisons [3, 24] (fig. 2). By not considering the consequences of genome duplication in comparisons of entire genomes or genomic segments such as orthologous BACs, one will chronically underestimate synteny/colinearity. One may also tend to erroneously interpret deviations from synteny/colinearity as a reflection of genomic fluidity such as transposition, not realizing that they are explicable by loss or degradation of duplicated genes.

The consequences of paleopolyploidy are remarkably different in euchromatic and heterochromatic regions, respectively. In *Arabidopsis* and *Oryza*, parallel arrangements of genes duplicated in the most recent whole-genome duplication event(s) cover 70–90% of the respective genomes. Nearly all of the regions in which duplication cannot be inferred are pericentromeric [3, 55] and/or heterochromatic [55], closely paralleling the chromosomal distribution of both conserved microsynteny, and recombination [55]. Preferential conservation of microsynteny in recombinogenic regions may suggest that gene rearrangement is generally deleterious, usually being eliminated from such regions by natural selection. In contrast, restructuring of relatively less recombinogenic centromeric regions may accelerate the transition to diploid inheritance and allow more rapid allele frequency changes and reduced genetic load [55].

The lower gene densities of pericentromeric and/or heterochromatic regions reduce power to discern both ancient duplication and conserved microsynteny. However, evidence from levels of divergence between duplicated rice genes reveals that there has also been extensive rearrangement of these regions that began shortly after paleopolyploidization and lasted until about 16 MYA. This appears to have largely involved migration of DNA between pericentromeric regions of different chromosomes [55].

Rapid progress in angiosperm genome sequencing offers growing scope to discern common features of the rare GD events that have led to successful lineages, providing clues about which, if any, of the many genome-wide consequences of polyploid formation might be connected to adaptation.

Spectrum of Fates of Duplicated Genes

Retention/loss of duplicated gene copies is not random. Duplicated genes resulting from whole-genome duplication have longer life expectancies than single-gene duplicates [56] and tend to reduplicate in new GD events. By contrast, genes that are rendered singleton following a genome duplication tend to be repeatedly returned to singleton status during successive genome duplications [22, 57].

Several studies have suggested that specific GO-based functional categories of genes may preferentially retain or lose copies [58–60]. While informative, the relatively broad GO-based classifications have the possibility of masking contrasting patterns in readily-distinguishable subgroups. For example, among protein-protein interaction domains, an abundant one (LRR) that is almost invariably retained in duplicate masks less-abundant ones (SET, TPR: [61]) that are usually returned to singleton status.

To investigate the degree of heterogeneity in patterns of gene retention/loss among GO categories, and also to provide a means to compare these patterns across broad taxonomic distances for which orthology could usually not be established, we have explored the tendencies for individual protein functional (Pfam) domains to occur among duplicates or singleton genes resulting from independent GD events in *Arabidopsis, Oryza, Tetraodon*, and *Saccharomyces*.

While frequencies of retention for the vast majority of duplicated genes in each taxon could not be distinguished from randomness, in angiosperms each of the two tails of the distribution were larger than would be expected by chance [61]. In other words, in angiosperms we find both non-random patterns of gene retention ('deletion-resistance') and non-random patterns of gene loss ('duplication-resistance'). In *Tetraodon* and *Saccharomyces* we were able to easily distinguish non-random patterns of gene retention – however the high overall level of gene loss made it difficult to distinguish exceptional cases. For example, 80% of *Tetraodon* domains were in singletons. While additional 'duplication-resistant' domains may have been found if one could have studied *Tetraodon* sooner after duplication, we could discern only one case still remaining – PF0400 (WD domain, G-beta repeat), with 78 occurrences in singletons and 3 in duplicates – and which was also duplication-resistant in rice. In yeast, two-thirds of domains were in singletons, and no significantly singleton-enriched domains were detected. However, against the generally low levels of duplicate gene retention, 22 and 9 domains were enriched in duplicated *Tetraodon* and yeast genes (respectively), versus only 4 in plants.

Those gene/domain families most frequently retained in duplicate in angiosperms and yeast were members of large heterogeneous families with wide spectra of effects. Indeed, the most duplicate-enriched domain was the same in *Tetraodon*, yeast, and both angiosperms (PF0069, protein kinase), and is also the most abundant domain in the rice genome. However, in *Tetraodon*, generally smaller domain families were preferentially retained in duplicate. Moreover, despite a general correlation across taxa in patterns of retention of domain-containing genes, preferential expansion of some domain families occurred in particular lineages. For example, Myb-like domains (PF00249) are dramatically expanded and largely duplicated in the angiosperms but essentially randomly distributed between yeast singletons and duplicates. *Tetraodon* has many domain types preferentially retained in duplicate, some related to animal-specific functions (motor/muscle function, secretin receptor) that show random distribution in plants. Analysis of additional duplications, as well as comparison of divergent taxa affected by common duplications, may reveal further lineage-specific trends that may have contributed to biological diversity, such as rapid growth and diversification of plant-specific AP2 gene families [61].

Many domains which occur too few times to reach statistical significance may also be under some degree of selection for singleton versus duplicate status. We calculated for each gene functional group (domain) the percentage of the total number of occurrences that were in singleton genes, thus removing bias due to differences in abundance of various domains. These fractions were closely correlated ($r = 0.59$, $p < 0.001$, $n = 1{,}006$) in *Arabidopsis* and *Oryza*, with lesser but still highly-significant correlations of the angiosperms to yeast ($r = 0.30$, $p < 0.001$; and $r = 0.39$, $p < 0.001$, respectively), and to *Tetraodon* (each $r = 0.29$, $p < 0.001$) [61].

A tantalizing notion for which there presently exist too little data to test, is whether speciation-causing gene loss [51–53] might be preferentially associated with duplication-resistant genes. If such an association exists, one would predict that the affected gene pairs would tend to show loss of different duplicate copies (i.e. divergent resolution) in different populations/species, leading to departures from conserved synteny.

Evolution of Duplicated Genes

A growing body of evidence points to seeming contradictions in our understanding of the evolution of duplicated genes. Classical views suggest that genome duplication is potentially advantageous as a source of genes with new functions [62–64]. However, such functions take time to evolve. Analysis of whole genome sequences shows that most genes are restored to singleton status following GD, reducing the possibility of such adaptive divergence. Further, expression patterns of duplicated genes diverge rapidly but their coding sequences diverge slowly [22, 65, 66]. Even in contemporary populations and among recently-formed subspecies, single nucleotide polymorphisms tend to encode fewer and less radical amino acid changes in genes for which there exists a duplicated copy at a 'paleologous' locus, than in 'singleton' genes [22, 67, 68].

Several avenues may offer some reconciliation between the conservative evolution of duplicated coding sequences and the classical 'functional divergence' model for duplicated gene evolution [62–64]. The greater retention in duplicate of long and complex proteins, together with their lower tolerance of non-synonymous mutations, suggest that the potential advantage conferred by the possibility of future neofunctionalization may often be outweighed by the immediate benefits of ensuring that the functions of essential genes are met even in the event of loss of function of one copy [22]. In no way does this preclude the evolution of unique functionality as one outcome in a spectrum of possibilities. Indeed, genes for which divergence was advantageous may have done it long ago, and are no longer recognizable as duplicates. Finally, expression divergence may in some instances turn out to be adaptive.

Functional buffering as a result of genome duplication may contribute, along with recombination, to reduced accumulation of degenerative mutations via Müller's ratchet [69]. Many predominantly clonally propagated angiosperms (such as banana and sugarcane), and also apomicts [70], are recently-formed polyploids – protection of critical functionality by duplication and preservation of essential genes would provide a mechanism by which, in the absence of recombination, they might avoid degeneration [69]. Similarly, repair processes fostered by whole-genome duplication are also key to the extraordinary radiation tolerance of *Deinococcus radiodurans* [71].

Angiosperms as a Facile Model

Looking forward, it appears likely that the angiosperms will play a growing role in clarifying the structural, functional and evolutionary consequences of paleopolyploidy. While the fully-sequenced genomes of several higher eukaryotes reveal evidence of GD in their evolutionary histories, the 'signal' from such events is faint – segmental duplication can only be resolved over 1–5% of the genomes of several mammals [43–47]. In contrast, angiosperms offer access to a large and growing number of independent genome duplications, at least some of which are of an age that permits us to discern both non-random patterns of gene retention and non-random patterns of gene loss. Population genetic theory predicts that the consequences of genome duplication in organisms with relatively small effective population sizes, such as angiosperms, may be more representative of other higher eukaryotes [52, 72] than would be (albeit valuable) new information about microbes such as yeast [73, 74].

Extensive sets of genomics tools and germplasm resources (in particular for many angiosperm taxa that include major crops), and the facility with which designed crosses and transgenic stocks can be made and studied, provide for a bridge between in silico inferences and in vivo functional studies. For example, a gene predicted in silico to be duplication-resistant and thus restored to singleton status soon after paleopolyploid formation, should be found in only a single copy within all members of the species. Introduction of a second copy and careful inspection of the phenotype might suggest the reason that duplication of that particular gene was maladaptive.

References

1 Galitski T, Saldanha AJ, Styles CA, Lander ES, Fink GR: Ploidy regulation of gene expression. Science 1999;285:251–254.

2 Hughes TR, Roberts CJ, Dai H, Jones AR, Meyer MR: Widespread aneuploidy revealed by DNA microarray expression profiling. Nat Genet 2000;25:333–337.

3 Bowers JE, Chapman BA, Rong J, Paterson AH: Unravelling angiosperm genome evolution by phylogenetic analysis of chromosomal duplication events. Nature 2003;422:433–438.

4 Wendel J, Stuber CW, Edwards MD, Goodman MM: Duplicated chromosome segments in *Zea mays* L.: Further evidence from hexokinase enzymes. Theor Appl Genet 1986;72:178–185.

5 Soltis DE, Soltis PS: Molecular-data and the dynamic nature of polyploidy. Crit Rev Plant Sci 1993;12:243–273.

6 Chittenden LM, Schertz KF, Lin Y-R, Wing RA, Paterson AH: A detailed RFLP map of *Sorghum bicolor* × *S. propinquum*, suitable for high-density mapping, suggests ancestral duplication of *Sorghum* chromosomes or chromosomal segments. Theor Appl Genet 1994;87:925–933.

7 Kishimoto N, Higo H, Abe K, Arai S, Saito A, Higo K: Identification of the duplicated segments in rice chromosomes 1 and 5 by linkage analysis of cDNA markers of known functions. Theor Appl Genet 1994;88:722–726.

8 Kowalski SP, Lan TH, Feldmann KA, Paterson AH: Comparative mapping of *Arabidopsis thaliana* and *Brassica oleracea* chromosomes reveals islands of conserved organization. Genetics 1994;138:499–510.

9 Nagamura Y, Inoue T, Antonio BA, Shimano T, Kajiya H, et al: Conservation of duplicated segments between rice chromosomes 11 and 12. Breed Sci 1995;45:373–376.

10 Paterson AH, Lan TH, Reischmann KP, Chang C, Lin YR, et al: Toward a unified genetic map of higher plants, transcending the monocot-dicot divergence. Nat Genet 1996;14:380–382.

11 Paterson AH, Bowers JE, Peterson DG, Estill JC, Chapman BA: Structure and evolution of cereal genomes. Curr Opin Genet Dev 2003;13:644–650.
12 Tuskan GA, Difazio S, Jansson S, Bohlmann J, Grigoriev I, et al: The genome of black cottonwood, *Populus trichocarpa* (Torr. & Gray). Science 2006;313:1596–1604.
13 Blanc G, Hokamp K, Wolfe KH: A recent polyploidy superimposed on older large-scale duplications in the *Arabidopsis* genome. Genome Res 2003;13:137–144.
14 Cui L, Wall PK, Leebens-Mack JH, Lindsay BG, Soltis DE, et al: Widespread genome duplications throughout the history of flowering plants. Genome Res 2006;16:738–749.
15 Pfeil BE, Schlueter JA, Shoemaker RC, Doyle JJ, et al: Placing paleopolyploidy in relation to taxon divergence: A phylogenetic analysis in legumes using 39 gene families. Syst Biol 2005;54:441–454.
16 Van de Peer Y: Computational approaches to unveiling ancient genome duplications. Nat Rev Genet 2004;5:752–763.
17 Roth J, Dilcher DL: Investigations of angiosperms from the eocene of North America: stipulate leaves of the Rubiaceae including a probable polyploid population. Am J Bot 1979;66: 1194–1207.
18 Masterson J: Stomatal size in fossil plants – evidence for polyploidy in majority of angiosperms. Science 1994;264:421–424.
19 Vision TJ, Brown DG, Tanksley SD: The origins of genomic duplications in *Arabidopsis*. Science 2000;290:2114–2117.
20 Zhang LQ, Vision TJ, Gaut BS: Patterns of nucleotide substitution among simultaneously duplicated gene pairs in *Arabidopsis thaliana*. Mol Biol Evol 2002;19:1464–1473.
21 Gao LZ, Innan H: Very low gene duplication rate in the yeast genome. Science 2004;306: 1367–1370.
22 Chapman BA, Bowers JE, Feltus FA, Paterson AH: Buffering crucial functions by paleologous duplicated genes may contribute cyclicality to angiosperm genome duplication. Proc Natl Acad Sci USA 2006;103:2730–2735.
23 Chapman BA, Bowers JE, Schulze SR, Paterson AH: A comparative phylogenetic approach for dating whole genome duplication events. Bioinformatics 2004;20:180–185.
24 Kellogg EA: It's all relative. Nature 2003;422:383–384.
25 Ming R, Liu SC, Lin YR, da Silva J, Wilson W, et al: Detailed alignment of *Saccharum* and *Sorghum* chromosomes: Comparative organization of closely related diploid and polyploid genomes. Genetics 1998;150:1663–1682.
26 Swigonova Z, Lai J, Ma J, Ramakrishna W, Llaca V, et al: On the tetraploid origin of the maize genome. Comp Funct Genomics 2004;5:281–284.
27 Kashkush K, Feldman M, Levy AA: Gene loss, silencing and activation in a newly synthesized wheat allotetraploid. Genetics 2002;160:1651–1659.
28 Ozkan H, Levy AA, Feldman M: Allopolyploidy-induced rapid genome evolution in the wheat (*Aegilops-Triticum*) group. Plant Cell 2001;13:1735–1747.
29 Shaked H, Kashkush K, Ozkan H, Feldman M, Levy AA: Sequence elimination and cytosine methylation are rapid and reproducible responses of the genome to wide hybridization and allopolyploidy in wheat. Plant Cell 2001;13:1749–1759.
30 Feldman M, Liu B, Segal G, Abbo S, Levy AA, Vega JM: Rapid elimination of low-copy DNA sequences in polyploid wheat: A possible mechanism for differentiation of homoeologous chromosomes. Genetics 1997;147:1381–1387.
31 Song KM, Lu P, Tang K, Osborn TC: Rapid genome change in synthetic polyploids of *Brassica* and its implications for polyploid evolution. Proc Natl Acad Sci USA 1995;92:7719–7723.
32 Ozkan H, Levy AA, Feldman M: Rapid differentiation of homeologous chromosomes in newly-formed allopolyploid wheat. Isr J Plant Sci 2002;50:S65–S76.
33 Kashkush K, Feldman M, Levy AA: Transcriptional activation of retrotransposons alters the expression of adjacent genes in wheat. Nat Genet 2003;33:102–106.
34 O'Neill RJW, O'Neill MJ, Graves JAM: Undermethylation associated with retroelement activation and chromosome remodelling in an interspecific mammalian hybrid. Nature 2002;420:106.
35 Chen ZJ, Pikaard CS: Epigenetic silencing of RNA polymerase I transcription: a role for DNA methylation and histone modification in nucleolar dominance. Genes Dev 1997;11:2124–2136.

36 Chen ZJ, Pikaard CS: Transcriptional analysis of nucleolar dominance in polyploid plants: Biased expression/silencing of progenitor rRNA genes is developmentally regulated in *Brassica*. Proc Natl Acad Sci USA 1997;94:3442–3447.
37 Comai L, Tyagi AP, Winter K, Holmes-Davis R, Reynolds SH, et al: Phenotypic instability and rapid gene silencing in newly formed *Arabidopsis* allotetraploids. Plant Cell 2000;12:1551–1568.
38 Lee HS, Chen ZJ: Protein-coding genes are epigenetically regulated in *Arabidopsis* polyploids. Proc Natl Acad Sci USA 2001;98:6753–6758.
39 Adams KL, Cronn R, Percifield R, Wendel JF: Genes duplicated by polyploidy show unequal contributions to the transcriptome and organ-specific reciprocal silencing. Proc Natl Acad Sci USA 2003;100:4649–4654.
40 Adams KL, Percifield R, Wendel JF: Organ-specific silencing of duplicated genes in a newly synthesized cotton allotetraploid. Genetics 2004;168:2217–2226.
41 Osborn TC, Quijada PA, Schranz ME, Pires JC, Zhao JW, et al: Flowering time divergence and genomic rearrangements in resynthesized *Brassica* polyploids (Brassicaceae). Biol J Linn Soc Lond 2004;82:675–688.
42 Rana D, van den Boogaart T, O'Neill CM, Hynes L, Bent E, et al: Conservation of the microstructure of genome segments in *Brassica napus* and its diploid relatives. Plant J 2004;40:725–733.
43 Bailey JA, Gu Z, Clark RA, Reinert K, Samonte RV, et al: Recent segmental duplications in the human genome. Science 2002;297:1003–1007.
44 Cheung J, Estivill X, Khaja R, MacDonald JR, Lau K, et al: Genome-wide detection of segmental duplications and potential assembly errors in the human genome sequence. Genome Biol 2003;4:R25.
45 Cheung J, Wilson MD, Zhang J, Khaja R, MacDonald JR, et al: Recent segmental and gene duplications in the mouse genome. Genome Biol 2003;4:R47.
46 Tuzun E, Bailey JA, Eichler EE: Recent segmental duplications in the working draft assembly of the brown Norway rat. Genome Res 2004;14:493–506.
47 Bailey JA, Church DM, Ventura M, Rocchi M, Eichler EE, et al: Analysis of segmental duplications and genome assembly in the mouse. Genome Res 2004;14:789–801.
48 Jaillon O, Aury JM, Brunet F, Petit JL, Stange-Thomann N, et al: Genome duplication in the teleost fish *Tetraodon nigroviridis* reveals the early vertebrate proto-karyotype. Nature 2004;431:946–957.
49 Van de Peer Y: *Tetraodon* genome confirms *Takifugu* findings: most fish are ancient polyploids. Genome Biol 2004;5:250.
50 Paterson AH, Bowers JE, Chapman BA: Ancient polyploidization predating divergence of the cereals, and its consequences for comparative genomics. Proc Natl Acad Sci USA 2004;101: 9903–9908.
51 Lynch M, Force AG: The origin of interspecific genomic incompatibility via gene duplication. Am Nat 2000;156:590–605.
52 Lynch M, O'Hely M, Walsh B, Force A: The probability of preservation of a newly arisen gene duplicate. Genetics 2001;159:1789–1804.
53 Scannell DR, Byrne KP, Gordon JL, Wong S, Wolfe KH: Multiple rounds of speciation associated with reciprocal gene loss in polyploid yeasts. Nature 2006;440:341–345.
54 Paterson AH, Freeling M, Sasaki T: Grains of knowledge: Genomics of model cereals. Genome Res 2005;15:1643–1650.
55 Bowers JE, Arias MA, Asher R, Avise JA, Ball RT, et al: Comparative physical mapping links conservation of microsynteny to chromosome structure and recombination in grasses. Proc Natl Acad Sci USA 2005;102:13206–13211.
56 Lynch M, Conery JS: The evolutionary fate and consequences of duplicate genes. Science 2000;290:1151–1155.
57 Seoighe C, Gehring C: Genome duplication led to highly selective expansion of the *Arabidopsis thaliana* proteome. Trends Genet 2004;20:461–464.
58 Maere S, De Bodt S, Raes J, Casneuf T, Van Montagu M, et al: Modeling gene and genome duplications in eukaryotes. Proc Natl Acad Sci USA 2005;102:5454–5459.
59 Blanc G, Wolfe KH: Functional divergence of duplicated genes formed by polyploidy during *Arabidopsis* evolution. Plant Cell 2004;16:1679–1691.

60 Blomme T, Vandepoele K, De Bodt S, Simillion C, Maere S, Van de Peer Y, et al: The gain and loss of genes during 600 million years of vertebrate evolution. Genome Biol 2006;7:R43.
61 Paterson AH, Chapman BA, Kissinger J, Bowers JE, Feltus FA, et al: Convergent retention or loss of gene/domain families following independent whole-genome duplication events in *Arabidopsis, Oryza, Saccharomyces,* and *Tetraodon*. Trends Genet 2006;22:597–602.
62 Stephens S: Possible significance of duplications in evolution. Adv Genet 1951;4:247–265.
63 Ohno S: Evolution by Gene Duplication (Springer, Berlin 1970).
64 Taylor JS, Raes J: Duplication and divergence: The evolution of new genes and old ideas. Annu Rev Genet 2004;38:615–643.
65 Jordan IK, Marino-Ramirez L, Koonin EV: Evolutionary significance of gene expression divergence. Gene 2005;345:119–126.
66 Davis JC, Petrov DA: Preferential duplication of conserved proteins in eukaryotic genomes. PLoS Biol 2004;2:e55.
67 Hughes MK, Hughes AL: Evolution of genes in a tetraploid animal, *Xenopus laevis*. Mol Biol Evol 1993;10:1360–1369.
68 Moore RC, Purugganan MD: The early stages of duplicate gene evolution. Proc Natl Acad Sci USA 2003;100:15682–15687.
69 Muller HJ: Some genetic aspects of sex. Am Nat 1932;66:118–138.
70 Bayer RJ, Stebbins GL: Chromosome-Numbers, Patterns of Distribution, and Apomixis in *Antennaria* (Asteraceae, Inuleae). Syst Bot 1987;12:305–319.
71 Levin-Zaidman S, Englander J, Shimoni E, Sharma AK, Minton KW, Minsky A: Ringlike structure of the *Deinococcus radiodurans* genome: A key to radioresistance? Science 2003;299:254–256.
72 Lynch M: The origins of eukaryotic gene structure. Mol Biol Evol 2006;23:450–468.
73 Gu Z, Steinmetz LM, Gu X, Scharfe C, Davis RW, Li WH: Role of duplicate genes in genetic robustness against null mutations. Nature 2003;421:63–66.
74 Christoffels A, Koh EG, Chia JM, Brenner S, Aparicio S, Venkatesh B: Fugu genome analysis provides evidence for a whole-genome duplication early during the evolution of ray-finned fishes. Mol Biol Evol 2004;21:1146–1151.
75 Rong J, Bowers JE, Schulze SR, Waghmare VN, Rogers CJ, et al: Comparative genomics of *Gossypium* and *Arabidopsis*: Unraveling the consequences of both ancient and recent polyploidy. Genome Res 2005;15:1198–1210.
76 Blanc G, Wolfe KH: Widespread paleopolyploidy in model plant species inferred from age distributions of duplicate genes. Plant Cell 2004;16:1667–1678.

Andrew H. Paterson
Plant Genome Mapping Laboratory, University of Georgia
111 Riverbend Road, Rm 228
Athens GA 30602 (USA)
Tel. +1 706 583 0162, Fax +1 706 583 0160, E-Mail paterson@uga.edu

Volff J-N (ed): Plant Genomes.
Genome Dyn. Basel, Karger, 2008, vol 4, pp 13–24

Genomic History and Gene Family Evolution in Angiosperms: Challenges and Opportunities

J. Sampedro, D. Cosgrove

Department of Biology, Pennsylvania State University, University Park, State College, Pa., USA

Abstract

Whole genome duplications (WGD) have been a frequent occurrence during the evolution of angiosperms, providing all gene families the opportunity to grow and diversify. Most of this potential growth has not been realized, since each WGD has been followed by massive gene losses. The likelihood of survival of gene duplicates after a WGD has been shown to depend on their function, as is also the case for single gene duplications. These two modes of growth have different functional and evolutionary implications and have had a markedly divergent impact on the evolution of different gene families. Despite duplications, gene losses, and translocations it is still possible in many cases to reconstruct the history of angiosperm genomic segments, sometimes back to the last ancestor of monocots and eudicots. This segmental phylogeny can in turn shed light on the evolution of the genes that form part of those segments. Position-based phylogeny can improve the resolution and correct artifacts created by phylogenies based on gene sequences, although a number of questions need to be resolved for its full potential to be fulfilled.

The sequencing of the entire genomes of *Arabidopsis*, rice and *Populus* [1–3] has revealed that whole genome duplications (WGD) have played a major role in the evolution of angiosperm gene families. In the case of the *Arabidopsis* genome, the pattern of duplicated segments has been interpreted to be the result of at least two and most likely three WGD events [4–6]. There is also agreement that at least one WGD is shared between rice and most other grasses [7, 8]. In the case of *Populus*, two polyploidies have been proposed, the most recent shared with *Salix* and the older one close to the divergence between the *Populus* and *Arabidopsis* lineages [3]. Similarly, the analysis of the partial genome sequences of *Medicago* and *Lotus* indicates the presence of a shared WGD, dated after the divergence between their lineage and that of *Populus* [9]. Finally, as a result of the analysis of large sets of cDNA sequences, additional WGD events have been proposed in the asterids lineage, as well as in basal eudicots, basal monocots or magnoliids [10, 11]. Although these recurring duplications of the entire genome can make it difficult to identify functional orthologs between distant species, they can also provide information on the timing and order of gene duplications, helping to clarify the phylogeny of gene families.

Gene Family Evolution in a Genomic Context

A model of the genome evolution in the *Arabidopsis* lineage suggests that the number of genes increased from 14,800 to 27,500 in the last 350 my and that approximately 60% of the new genes that have survived to the present were created by a WGD [5]. This increase is the final result of a succession of three polyploidizations, each followed by massive gene losses. It has been estimated that only one in six of the new genes created by the most recent polyploidy in the *Arabidopsis* lineage has survived to the present [5, 12]. The partial sequencing of the maize genome has shown that only 4.8 my after a WGD, already more than half of the extra genes have disappeared [13]. A similarly fast rate of loss has been described for a WGD in yeasts [14].

The loss of genes after a WGD is not a completely random process. A number of studies have shown that genes with certain functions, such as transcription factors, have a significantly higher probability of survival, while other genes appear to show the opposite effect [5, 12, 15]. This effect is also evident at the scale of individual gene lineages, since duplicates retained after a WGD are more likely to successfully duplicate again in latter polyploidies [15, 16]. Another key source of new genes is the process of tandem duplication, which explains at least half of the recently duplicated genes in *Arabidopsis* and somewhat less in rice [10, 17]. As with those created by a WGD, tandem duplicates are also more likely to survive in some families than others [17].

An interesting observation of genome-wide surveys is that for some families the survival rate of duplicated genes seems to depend strongly on the mechanism that created them [5, 17, 18]. Transcription factors, for example, are more likely to survive in duplicate after a WGD, possibly because single gene duplications in a regulatory pathway are likely to cause deleterious gene dosage effects [19]. The duplication of the entire genome could thus create a unique opportunity for the duplication of complete regulatory networks and the consequent increase in developmental complexity [12].

The physical distance between the duplicates is also very different between tandem duplications and a WGD and this can have important consequences for the subsequent pattern of evolution. Closely spaced duplicates are more likely to go through homogenization processes like concerted or birth-and-death evolution [20]. The long term survival of duplicated genes, whatever their origin, is believed to require a differentiation of functions either through neofunctionalization of one of the copies, the partition of the original functions between the two genes, or a combination of the two [21]. Subfunctionalization, the most likely survival mechanism in the short term, could free one or the two copies to later adopt new functions [22]. After a WGD it is common to find unequal rates of evolution between gene duplicates, as was shown in the well-studied case of yeasts [23, 24]. In *Arabidopsis* there is also evidence of this process for the most recent polyploidy [12]. However, unequal rates of evolution cannot be interpreted as evidence of neofunctionalization, since subfunctionalization could also create gene pairs with different degrees of functional constraint.

Position-Based Phylogeny

Genomic sequencing not only provides a complete sampling of the size and diversity of gene families, it also shows the localization of genes and their genomic context, and this information can provide important evidence regarding the evolution of a gene family. Colinearity between the genomes of two species, for example, can be used to identify orthologous genes with more confidence, as has been recently attempted for the entire *Arabidopsis* and *Populus* genomes [3]. The important role of WGD in the growth of many angiosperm gene families can also be exploited by linking individual gene duplications to these large-scale events. This has already been done, on an individual basis, for a number of families [25–27]. Software that automates the process of associating gene and segmental duplications has also been used to analyze the growth of 50 *Arabidopsis* gene families [18].

Going one step further, the conservation of synteny between and within genomes can potentially be used to reconstruct ancestral genomic segments in the common ancestor of one or more species, as well as reconstruct the history of duplications and speciation events that resulted in the present-day situation [28]. This segmental phylogeny can then be used to clarify the pattern and timing of individual gene duplications, a process that could be called 'position-based' phylogeny. This idea has been applied to the study of expansins, a superfamily of cell wall proteins, in order to complement or even correct 'sequence-based' phylogenetic trees [28, 29].

Position-based phylogeny can be particularly useful as a way of identifying groups of orthologous genes between distantly related species, such as *Arabidopsis* and rice. In the case of the expansins, as well as many other gene families, it is difficult to identify orthologous groups with any degree of confidence using standard phylogenetic methods [28, 30]. An additional problem for trees that include monocot and eudicot sequences is the nucleotide and amino acid compositional biases found in grasses [31], and which are also evident for the rice expansins [28]. Only by combining position-based and sequence-based phylogeny, was it possible to classify the 94 expansins in *Arabidopsis* and rice into 17 orthologous groups or clades [28], a conclusion that was later confirmed and strengthened by the analysis of the *Populus* genome [29].

The use of position-based phylogeny for *Arabidopsis* and rice genes depends on the possibility of detecting colinearity of gene order between both species, a task made more difficult by the polyploidies that took place in these lineages. After a single WGD, assuming 16% of duplicated genes survive as such [12], the parent segment would only share 58% of its genes on average with each descendant. A recent study that compared directly pairs of *Arabidopsis* and rice segments found that the regions where significant synteny can still be detected encompass only 33% of the *Arabidopsis* genes and 17% of those from rice [32]. A more promising approach is to try to assemble the several *Arabidopsis* and rice genomic fragments that descend from a common ancestor, and then judge the colinearity of the whole set. However, an automated application of this methodology to the entire genomes of both species had only limited success, with the colinear regions covering only 30% and 14% of the *Arabidopsis* and rice genomes [33]. In a similar way, in the study of the expansin superfamily, sets of up to 10 rice and *Arabidopsis* homologous segments were aligned, but with this small-scale approach microsynteny was found for all 12 cases where orthologous expansin genes are still present in both species [28], suggesting that the extent of colinearity may have been underestimated in previous studies.

The History of a Segment

An example can serve to illustrate the possibilities and difficulties of position-based phylogeny. One of the expansin clades where position-based and sequence-based phylogenies do not agree is EXPA-IV, with one member in rice and five each in *Arabidopsis* and *Populus* [28, 29]. All the genomic segments where these genes are found, as well as some others that no longer contain expansins, are homologous to each other and can be traced to a single ancestral segment (fig. 1). Among the many gene families present in the ancestral segment, three other families were selected for detailed examination. All four families show a high survival rate after segmental duplications, so that a total of 13 *Dof* genes, 10 type-2C protein phosphatases (*PP2C*), 12 receptor-like cytoplasmic kinases (*RLCK*), and 11 *EXPA-IV* expansins are present in 16 homologous genomic segments, usually in the same order and orientation and always very close to each other (fig. 1a). These four families have also been the subject of genome-wide phylogenetic analyses, which are important for the identification of proper outgroups [28, 34–36].

The history of these segments was reconstructed on the basis of previously published analyses of WGD events on the three lineages [3, 4, 6, 7], as well as synteny results and sequence-based phylogenetic analyses [28, 29]. In addition to the three proposed polyploidies in the *Arabidopsis* lineage (labeled α, β and γ), and the WGD events specific to rice (ρ) and *Populus* (S), an additional segmental duplication was found, involving more than 200 genes and placed between the β and γ events (fig. 1a).

The sequence-based phylogenetic trees for the *Dof*, *PP2C* and *RLCK* families are all in accord with the segmental phylogeny, if only nodes supported by bootstrap values above 50% are taken into account (fig. 1b). While *PP2C* genes show a fairly strong phylogenetic signal, the analyses of both *Dof* and *RLCK* genes result in completely unresolved trees beyond the most recent nodes (fig. 1b). Since these genes appear to have experienced duplication and speciation simultaneously with the segments where they are found, it is likely that the true gene phylogeny for both families is coincident with the segmental phylogeny. At a minimum, the results from position-based phylogeny can be taken as a working hypothesis to direct further research.

In the case of the expansins, the sequence-based trees favor a different phylogeny from that of the surrounding segments (fig. 1b). The difference is mainly a question of rooting and there are several additional reasons to believe it could be caused by a long-branch attraction artifact, including high ratios of nonsynonymous to synonymous substitutions in some branches [28, 29]. This example shows how position-based phylogeny can be used to supplement weakly supported sequence-based trees and, moreover, to indicate the possible presence of phylogenetic artifacts. In addition, this kind of study also allows the quantification of gene deaths, by locating segments that no longer contain the genes present in their ancestors [28].

Genes Versus Segments

Because tandem duplications and WGD are the main sources of new genes for most angiosperm gene families and both processes preserve the

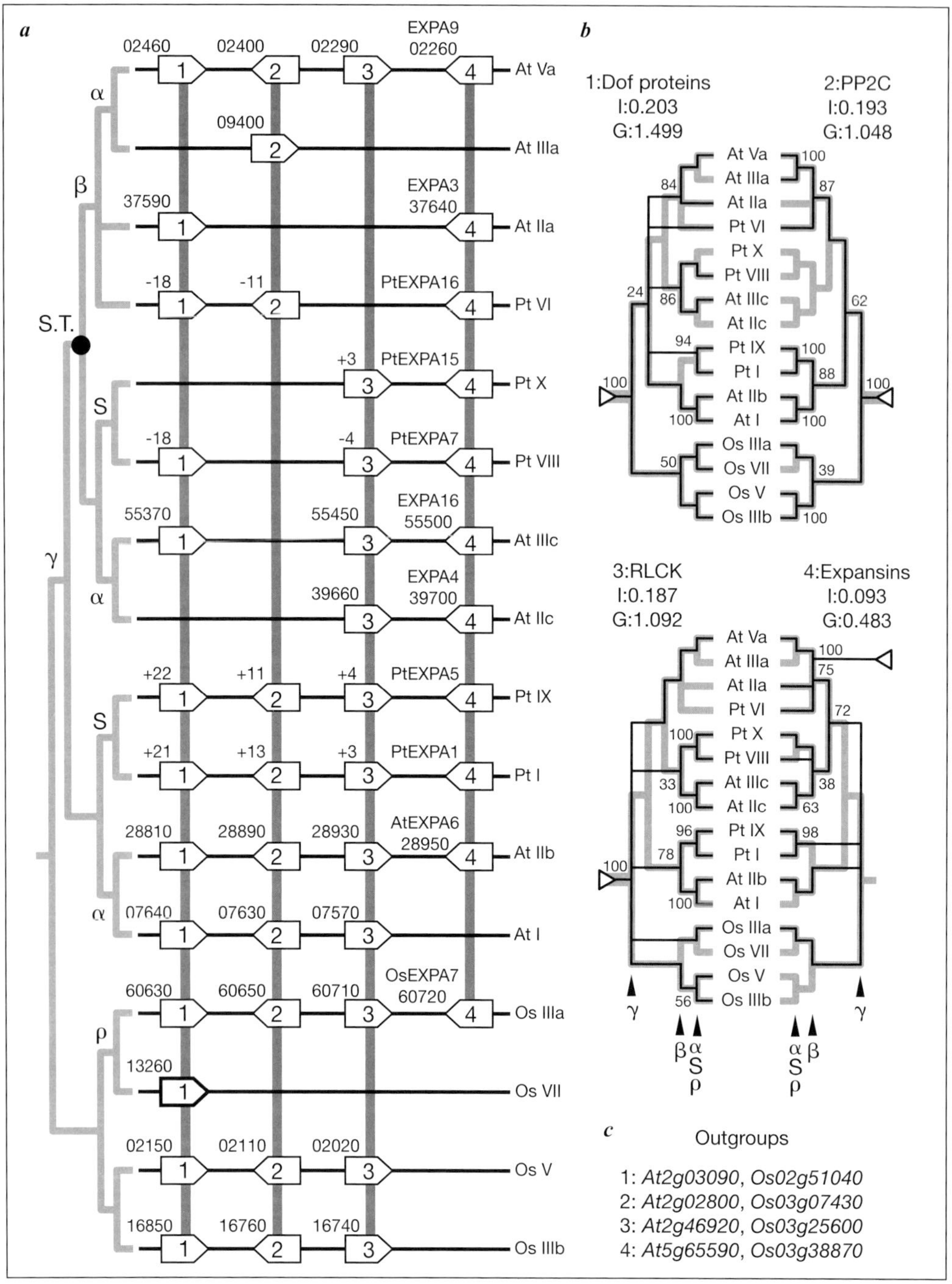
a
b
1:Dof proteins
I:0.203
G:1.499
2:PP2C
I:0.193
G:1.048
3:RLCK
I:0.187
G:1.092
4:Expansins
I:0.093
G:0.483
c
Outgroups
1: At2g03090, Os02g51040
2: At2g02800, Os03g07430
3: At2g46920, Os03g25600
4: At5g65590, Os03g38870

relative position of genes, the potential use of position-based phylogeny is extensive. However, there are a number of difficulties that should be taken into account. Widespread translocation, for example, can erase the usefulness of positional information, and plant lineages where this phenomenon is common will not be amenable to this methodology. In the case of the expansin superfamily, two translocation events were detected in the *Arabidopsis* lineage and six in that of rice, compared with 20 and 37 duplications, respectively, where the synteny was maintained [29]. For a family of resistance genes in *Arabidopsis*, 96% of gene duplications were also found to preserve synteny [37]. On the other hand, a recent study of two large genomic regions in maize has found that almost a third of the genes seem to have moved since this species diverged from rice [13].

It is also important to keep in mind that finding two homologous genes in colinear positions does not guarantee that their duplication time coincides with that of the segment, since there are at least two mechanisms that could make gene phylogeny different from segment phylogeny (fig. 2). In the first case, a gene conversion that takes place after a segmental duplication would cause position-based phylogeny to overestimate the age of the paralogous genes. Conversely, a preexisting tandem duplication followed by alternate gene losses would cause position-based phylogeny to underestimate the age of the gene duplication. These artifacts in position-based phylogenies are more likely when speciation and segmental duplication events are close in time to each other.

The prevalence of gene conversion between paralogs after a segmental duplication is hard to estimate, because direct evidence of this phenomenon can be detected only if it affects part of a gene or it is fairly recent. A study of 242 *Arabidopsis* gene pairs created in the most recent WGD could not find a single

Fig. 1. Position-based phylogeny. ***a*** Alignment of genomic segments from *Arabidopsis* (At), *Populus* (Pt) and rice (Os), identified by their chromosome number and an additional letter when necessary. Only genes from the *Dof* (1), *PP2C* (2), *RLCK* (3) and expansin families (4) are shown, identified by annotation numbers in *Arabidopsis* (TAIR) and rice (TIGR) or position relative to the expansin gene in *Populus*. A segmental phylogeny to the left identifies the nodes linked to a WGD, as well as a segmental duplication in tandem (S.T.). ***b*** Maximum-likelihood phylogenetic trees of the four gene families superimposed on the segmental phylogeny. Trees based on protein sequences were obtained with PHYML [45], using the values for gamma (G) and invariant sites (I) shown in the figure, as estimated by the program. Branches with bootstrap values (based on 200 replications) below 50% were collapsed when divergent from the segmental phylogeny. ***c*** Outgroups used to root the sequence-based phylogenies.

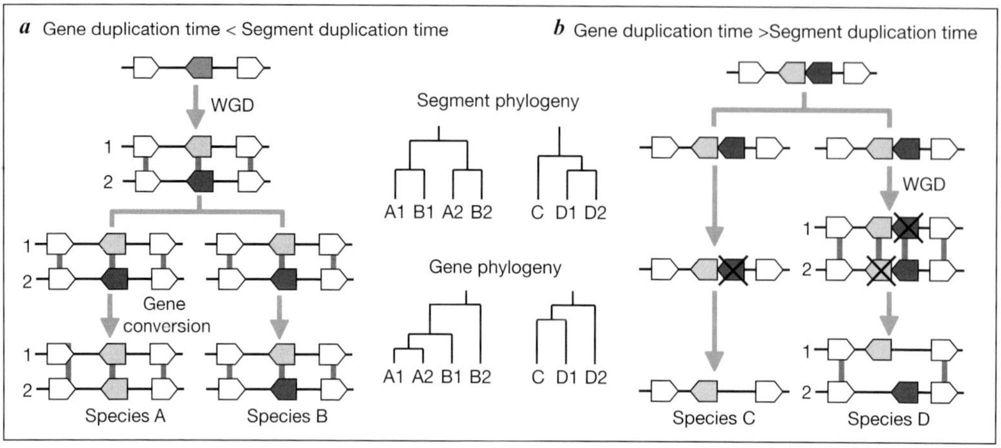

Fig. 2. Possible conflicts between segment and gene phylogeny. ***a*** Caused by a gene conversion. ***b*** Caused by a tandem duplication followed by asymmetric gene losses.

instance of gene conversion [38]. More recently, a few cases of possible conversion between distant genes have been found for *Arabidopsis* gene families where this phenomenon is expected to be particularly favored, but at much lower rates than between neighboring genes [37, 39]. On the other hand, some studies have suggested that gene conversion between paralogs may have been common for some time after the WGD in yeasts [24, 40]. However, some of these apparent conversions could be artifacts due to strong codon-usage bias and the remaining ones seem mostly limited to genes where sequence divergence is particularly slow [41]. Similarly, some studies have found evidence of frequent gene conversions between paralogs created by the most recent polyploidies in *Arabidopsis* and rice [16, 42], but codon-usage bias should be considered as an alternative explanation. This is clearly an important question that needs to be investigated further.

Regarding the potential distortions caused by tandems, this problem is more likely to be found in families where the birth-and-death mode of evolution is the norm [20], and for them position-based phylogeny may not be of much use. In the example discussed above none of the expansin genes appear in tandem (fig. 1), and to make sequence-based and position-based phylogenies consistent would require an ancient triple tandem followed by a complicated sequence of unilateral gene losses. Since conversions do not take place between species and gene losses occur independently in different lineages, both kinds of artifacts will be easier to detect as more genomic sequences become available and can be incorporated to the segmental phylogeny.

Dating Segmental Duplications

The likelihood of artifacts like those discussed above is hard to quantify and is likely to vary between different families. This problem makes it difficult to estimate the confidence one can put in position-based phylogeny. Another potential source of error is the segmental phylogeny itself, which depends on accurately dating segmental duplication and speciation events with respect to each other. Three different methods can be used to resolve the phylogeny of segments, each of them with its own advantages and problems [43]. Synonymous substitutions between paralogous and orthologous gene duplicates can be used as a proxy for age, and this approach has been used to characterize all the proposed angiosperm polyploidies [3, 4, 8–11]. A serious limitation of this methodology is the saturation of synonymous sites for very ancient duplications. In addition, unequal rates of substitution between gene pairs make it difficult to discriminate between closely related events.

An alternative approach is to construct phylogenetic trees for duplicated genes within the segments, a method that has also been frequently employed in plants [3, 6, 7, 9, 44]. Some caution is needed with this approach since, as already discussed, unequal evolutionary rates appear to be common in gene pairs created during a WGD. A recent study in yeast found that rate acceleration in one of the branches can frequently distort the resulting tree topology, creating long-branch attraction artifacts in up to 54% of trees [23]. This could also explain some contradictory results in the dating of plant polyploidies. On the basis of phylogenetic trees, the β event in the *Arabidopsis* lineage was dated close to the divergence between monocots and eudicots [44]. A more recent study, using the *Populus* genome and a combination of methods, strongly suggests that it took place much later, very close to the divergence between the *Populus* and *Arabidopsis* lineages [3].

Some problems of the phylogenetic method can be avoided by filtering out gene pairs that show evidence of unequal rates but, as with the use of synonymous substitutions, resolving closely spaced events will still be difficult. In the case of the β WGD and the split of the *Populus* and *Arabidopsis* lineages, a large number of phylogenetic trees were found that strongly supported each of the alternative topologies [3]. The even older γ WGD in *Arabidopsis* has been dated, on the basis of contradictory phylogenetic trees, before the split of eudicots and monocots [6, 44]. In the analysis of the expansin superfamily, a γ event specific to the eudicot lineage was found to fit better with the pattern of synteny [28].

In the example discussed above (fig. 1), the phylogeny of the *PP2C* family favors a late γ event, two other trees are unresolved and the only one that favors an early γ has signs of unequal rates of evolution.

The solution to the dating of the γ and β events could come from the use of genomic alignments. When *Kluyveromyces waltii* was fully sequenced, it was possible to align each segment in the genome to two segments in *Saccharomyces cerevisiae*, clearly showing that the WGD in the latter occurred after the split of the two lineages [24]. Even if a complete alignment is not possible, the degree of synteny conserved between two genomes with a shared WGD should be much higher than the synteny found within each of them, as has been shown for the partially sequenced genomes of *Medicago* and *Lotus* [9]. A WGD, unlike a speciation event, is expected to be followed by fast and massive gene losses that quickly disrupt the conservation of synteny [14]. In the case of the older angiosperm polyploidies, this kind of analysis is made more difficult due to the presence of later duplications, but it may still be the best approach for dating the γ event. If γ is shared between *Arabidopsis* and rice, it should be possible to detect it in the latter species and post-γ rice segments should align better to post-γ *Arabidopsis* segments than each of them aligns to their γ duplicates.

Despite the difficulties, the phylogeny of segments could potentially be resolved with more confidence than the phylogeny of the component genes, since many of the distortions that can affect the latter are averaged when analyzing the former. In addition, the argument from synteny is a source of information independent from the vagaries of gene sequence evolution. This is particularly true for a WGD, where the information of the entire genome can be brought to bear on the dating. The identification and dating of segmental duplications not linked to a WGD is more challenging, since the number of gene pairs that can be analyzed is more limited. When studying the expansin superfamily, we found two examples of independent segmental duplications, both in the form of tandemly repeated segments, which is likely to be the norm for these events [28].

Future Directions

As more angiosperm genomes are sequenced it will become easier to align them and to reconstruct, with a high degree of confidence, the history of the different regions from their last common ancestor to the present. Due to the important role of WGD events in the evolution of angiosperms, the detailed genomic history of this group would be a key tool to understand the growth and diversification of many gene families. It would also facilitate in great measure the correct identification of orthologs and paralogs, an essential task in the study of functional evolution.

References

1 Arabidopsis Genome Initiative: Analysis of the genome sequence of the flowering plant *Arabidopsis thaliana*. Nature 2000;408:796–815.

2 International Rice Genome Sequencing Project: The map-based sequence of the rice genome. Nature 2005;436:793–800.

3 Tuskan GA, DiFazio S, Jansson S, Bohlmann J, Grigoriev I, et al: The genome of black cottonwood, *Populus trichocarpa* (Torr. & Gray). Science 2006;313:1596–1604.

4 Simillion C, Vandepoele K, Van Montagu MC, Zabeau M, Van de Peer Y: The hidden duplication past of *Arabidopsis thaliana*. Proc Natl Acad Sci USA 2002;99:13627–13632.

5 Maere S, De Bodt S, Raes J, Casneuf T, Van MM, Kuiper M, Van de Peer Y: Modeling gene and genome duplications in eukaryotes. Proc Natl Acad Sci USA 2005;102:5454–5459.

6 Bowers JE, Chapman BA, Rong J, Paterson AH: Unravelling angiosperm genome evolution by phylogenetic analysis of chromosomal duplication events. Nature 2003;422:433–438.

7 Paterson AH, Bowers JE, Chapman BA: Ancient polyploidization predating divergence of the cereals, and its consequences for comparative genomics. Proc Natl Acad Sci USA 2004;101:9903–9908.

8 Yu J, Wang J, Lin W, Li S, Li H, et al: The genomes of *Oryza sativa*: a history of duplications. PLoS Biol 2005;3:e38.

9 Cannon SB, Sterck L, Rombauts S, Sato S, Cheung F, et al: Legume genome evolution viewed through the *Medicago truncatula* and *Lotus japonicus* genomes. Proc Natl Acad Sci USA 2006;103:14959–14964.

10 Blanc G, Wolfe KH: Widespread paleopolyploidy in model plant species inferred from age distributions of duplicate genes. Plant Cell 2004;16:1667–1678.

11 Cui L, Wall PK, Leebens-Mack JH, Lindsay BG, Soltis DE, et al: Widespread genome duplications throughout the history of flowering plants. Genome Res 2006;16:738–749.

12 Blanc G, Wolfe KH: Functional divergence of duplicated genes formed by polyploidy during *Arabidopsis* evolution. Plant Cell 2004;16:1679–1691.

13 Bruggmann R, Bharti AK, Gundlach H, Lai J, Young S, et al: Uneven chromosome contraction and expansion in the maize genome. Genome Res 2006;16:1241–1251.

14 Scannell DR, Byrne KP, Gordon JL, Wong S, Wolfe KH: Multiple rounds of speciation associated with reciprocal gene loss in polyploid yeasts. Nature 2006;440:341–345.

15 Seoighe C, Gehring C: Genome duplication led to highly selective expansion of the *Arabidopsis thaliana* proteome. Trends Genet 2004;20:461–464.

16 Chapman BA, Bowers JE, Feltus FA, Paterson AH: Buffering of crucial functions by paleologous duplicated genes may contribute cyclicality to angiosperm genome duplication. Proc Natl Acad Sci USA 2006;103:2730–2735.

17 Rizzon C, Ponger L, Gaut BS: Striking similarities in the genomic distribution of tandemly arrayed genes in *Arabidopsis* and rice. PLoS Comput Biol 2006;2:e115.

18 Cannon SB, Mitra A, Baumgarten A, Young ND, May G: The roles of segmental and tandem gene duplication in the evolution of large gene families in *Arabidopsis thaliana*. BMC Plant Biol 2004;4:10.

19 Veitia RA: Gene dosage balance in cellular pathways: implications for dominance and gene duplicability. Genetics 2004;168:569–574.

20 Nei M, Rooney AP: Concerted and birth-and-death evolution of multigene families. Annu Rev Genet 2005;39:121–152.

21 Force A, Lynch M, Pickett FB, Amores A, Yan YL, Postlethwait J: Preservation of duplicate genes by complementary, degenerative mutations. Genetics 1999;151:1531–1545.

22 He X, Zhang J: Rapid subfunctionalization accompanied by prolonged and substantial neofunctionalization in duplicate gene evolution. Genetics 2005;169:1157–1164.

23 Fares MA, Byrne KP, Wolfe KH: Rate asymmetry after genome duplication causes substantial long-branch attraction artifacts in the phylogeny of *Saccharomyces* species. Mol Biol Evol 2006;23:245–253.
24 Kellis M, Birren BW, Lander ES: Proof and evolutionary analysis of ancient genome duplication in the yeast *Saccharomyces cerevisiae*. Nature 2004;428:617–624.
25 Li X, Duan X, Jiang H, Sun Y, Tang Y, et al: Genome-wide analysis of basic/helix-loop-helix transcription factor family in rice and *Arabidopsis*. Plant Physiol 2006;141:1167–1184.
26 Kim J, Shiu SH, Thoma S, Li WH, Patterson SE: Patterns of expansion and expression divergence in the plant polygalacturonase gene family. Genome Biol 2006;7:R87.
27 Jiang D, Yin C, Yu A, Zhou X, Liang W, et al: Duplication and expression analysis of multicopy miRNA gene family members in *Arabidopsis* and rice. Cell Res 2006;16:507–518.
28 Sampedro J, Lee Y, Carey RE, dePamphilis C, Cosgrove DJ: Use of genomic history to improve phylogeny and understanding of births and deaths in a gene family. Plant J 2005;44:409–419.
29 Sampedro J, Carey RE, Cosgrove DJ: Genome histories clarify evolution of the expansin superfamily: new insights from the poplar genome and pine ESTs. J Plant Res 2006;119:11–21.
30 Li Y, Jones L, McQueen-Mason S: Expansins and cell growth. Curr Opin Plant Biol 2003;6: 603–610.
31 Wang HC, Singer GA, Hickey DA: Mutational bias affects protein evolution in flowering plants. Mol Biol Evol 2004;21:90–96.
32 Wang X, Shi X, Li Z, Zhu Q, Kong L, et al: Statistical inference of chromosomal homology based on gene colinearity and applications to *Arabidopsis* and rice. BMC Bioinformatics 2006;7:447.
33 Simillion C, Vandepoele K, Saeys Y, Van de Peer Y: Building genomic profiles for uncovering segmental homology in the twilight zone. Genome Res 2004;14:1095–1106.
34 Schweighofer A, Hirt H, Meskiene I: Plant PP2C phosphatases: emerging functions in stress signaling. Trends Plant Sci 2004;9:236–243.
35 Lijavetzky D, Carbonero P, Vicente-Carbajosa J: Genome-wide comparative phylogenetic analysis of the rice and *Arabidopsis* Dof gene families. BMC Evol Biol 2003;3:17.
36 Shiu SH, Karlowski WM, Pan R, Tzeng YH, Mayer KFX, Li WH: Comparative analysis of the receptor-like kinase family in *Arabidopsis* and rice. Plant Cell 2004;16:1220–1234.
37 Baumgarten A, Cannon S, Spangler R, May G: Genome-level evolution of resistance genes in *Arabidopsis thaliana*. Genetics 2003;165:309–319.
38 Zhang L, Vision TJ, Gaut BS: Patterns of nucleotide substitution among simultaneously duplicated gene pairs in *Arabidopsis thaliana*. Mol Biol Evol 2002;19:1464–1473.
39 Mondragon-Palomino M, Gaut BS: Gene conversion and the evolution of three leucine-rich repeat gene families in *Arabidopsis thaliana*. Mol Biol Evol 2005;22:2444–2456.
40 Sugino RP, Innan H: Estimating the time to the whole-genome duplication and the duration of concerted evolution via gene conversion in yeast. Genetics 2005;171:63–69.
41 Lin YS, Byrnes JK, Hwang JK, Li WH: Codon-usage bias versus gene conversion in the evolution of yeast duplicate genes. Proc Natl Acad Sci USA 2006;103:14412–14416.
42 Benovoy D, Morris RT, Morin A, Drouin G: Ectopic gene conversions increase the G + C content of duplicated yeast and *Arabidopsis* genes. Mol Biol Evol 2005;22:1865–1868.
43 de Peer YV: Computational approaches to unveiling ancient genome duplications. Nat Rev Genet 2004;5:752–763.
44 Chapman BA, Bowers JE, Schulze SR, Paterson AH: A comparative phylogenetic approach for dating whole genome duplication events. Bioinformatics 2004;20:180–185.
45 Stéphane G, Gascuel O: A simple, fast, and accurate algorithm to estimate large phylogenies by maximum likelihood. Syst Biol 2003;52:696–704.

Javier Sampedro
Department of Biology, 208 Mueller Lab
Pennsylvania State University, University Park
State College, PA 16802 (USA)
Tel. +1 814 865 3752, Fax +1 814 865 9131, E-Mail jus16@psu.edu

Volff J-N (ed): Plant Genomes.
Genome Dyn. Basel, Karger, 2008, vol 4, pp 25–40

The Evolutionary Position of Subfunctionalization, Downgraded

M. Freeling

Department of Plant and Microbial Biology, University of California, Berkeley, Calif., USA

Abstract

Current data from complete eukaryotic genomes indicate that ancestral gene duplications, followed by a mutational process called fractionation, generated profound and orderly changes in gene content. Most of these duplicated genes are removed. At least three hypotheses may explain the exceptional genes retained post-duplication: (1) *Gain-of-Function*; (2) *Subfunctionalization*, and (3) *Balanced Gene Drive*. Each is evaluated as an explanation for gene content data. Subfunctionalization, the most popular explanation, predicts no relationship at all between gene function and post-duplicate retention, and if there were particular sorts of 'subfunctionalizable' genes, these should be over-retained following any sort of duplication. Duplications may be local, segmental or whole genome. Gene content data from three plant genomes, reflecting three independent tetraploidies and many tandem duplications, are not explained by Subfunctionalization. Specifically, genes encoding transcription factors and ribosomal components are significantly over-retained following tetraploidy and under-retained among local duplicates. In addition, transcription factor families in *Arabidopsis* show a reciprocal relationship when retention is monitored after local duplication versus after tetraploidy; only Balanced Gene Drive predicts reciprocity. Vertebrates also retain genes nonrandomly following tetraploidies, but the data are preliminary. Removing subfunctionalization as the duplicate retention mechanism is of high theoretical importance. It clears the way for 'Mutationist' hypotheses that may help explain baffling adaptations and trends in eukaryotic evolution that have been largely ignored. This essay recognizes the potential evolutionary importance of saltatory chromosomal events that may change gene content – expand gene families – independent of allelic diversity.

'Lineage-specific expansions seem to be one of the principal means of adaptation and one of the most important sources of organizational and regulatory diversity in crown-group eukaryotes' [1].

'Subfunctionalization' is not only a big word, but is a big idea. Force and coworkers [2] originated this particularly useful Subfunctionalization idea to explain why so many of the gene pairs generated by ancient eukaryotic tetraploidies are retained as pairs today, and also why there are so many genes in local (tandem) arrays. The subfunctionalization idea is similar to a scheme advanced specifically for exon evolution [3]. A definition of Subfunctionalization is given below, along with definitions of two other ideas that have been put forward to explain duplicate retention: the classic Gain-of-Function model of E.B. Lewis [4] and especially S. Ohno [5], and the Balanced Gene Drive model [6] derived from the robust Gene Balance Hypothesis [7–10]. Subfunctionalization has proved to be particularly useful theoretically, probably because the mutations used are selectively neutral. The word 'subfunctionalization' is capitalized in this work when it connotes a model to explain over-retention following duplication rather than simply a divergence mechanism.

The Subfunctionalization explanation for duplicate gene retention occupies a particularly important position in evolutionary theory, a position that – were alternative or additional mechanisms actually at play – has the potential to 'drive' or lend direction to evolutionary trends that would otherwise defy explanation. That is because retention post-duplication, if biased, could mediate trends in specific gene or gene family expansion. For example, the trend of increasing maxima of morphological complexity in eukaryotes has been explained by replacing subfunctionalization with an alternative retention mechanism called Balanced Gene Drive [6]. A careful definition of 'morphological complexity', 'developmental boundaries' and 'drive' were necessary parts of this explanation. Explanation of trends and baffling adaptations constitutes one of evolutionary theory's biggest challenges, and one that the Modern Synthesis (Dobzhanski's consensus from the 1940s and 50s, [11]) does not address, almost as if there were no trends to explain, and no baffling adaptations. Among the many possible eukaryotic trends [12, 13], particularly remarkable are the increases in environmental endurance in plants, number of synapses in animals, and increases in the complexity of symbioses and inter-species dependencies in many lineages. Particularly baffling are adaptations that seem to have required the recruitment of modules (networks) of genes rather than depending on a series of single gene gains-of-function (e.g. the evolution of the feather).

Three Candidate Hypotheses as to Why Duplicate Genes Are Retained More Often than Expected

Gain-of-Function Hypothesis [4, 5]

Following duplication of any sort, selection for function for one or both duplicate genes is relaxed, and mutations are possible. A mutation in either gene could confer a novel function, which could then be positively selected. The result would be that this dominant allele would increase its frequency in the gene pool, and bring a stretch of chromosome with it. This is a one-hit, adaptive event. Synonym: neofunctionalization hypothesis.

Subfunctionalization Hypothesis [2]

Following duplication of any sort, complementary mutations in each duplicate gene are predicted to occur, assuming that the mutations are in different, dispensible regions of the gene. Thus, the entire ancestral function is now spread between two subfunctionalized genes, genes that must both be retained to maintain the status quo defined by the ancestor gene. This is a two-hit mutational event at minimum, but each mutation is selectively neutral. Synonym: duplication, degeneration, complementation (DDC) model.

Balanced Gene Drive Hypothesis [6]

The Gene Balance Hypothesis [7–10] provides the basis for predictions about what sort of genes should be retained following either a tetraploidy or an individual gene duplication. The principle is that some genes are dose-sensitive, so that too many or too few genes – in relation to the dosage of gene with which the gene must interact – lower fitness (haploinsufficiency or triploinsufficiency in relation to a diploid). Dosage-sensitive genes are retained only because they cannot be removed, and the status quo is maintained. This means that genes that function in particular ways should be over-retained following tetraploidy and should be under-retained following tandem duplication. Selection is immediate upon the duplication event. The mechanical basis for dosage-sensitivity is not completely understood, but may involve protein-protein interactions, length of a regulatory cascade and/or particular sorts of positive and negative regulation [8]. Genes that are dose sensitive by any of these various mechanisms have been called 'connected genes' [6], and are expected to be dose-sensitive. Retention is physiological; zero mutational hits are necessary. Synonym: gene dosage hypothesis.

Introduction to Gene and Genome Duplication

When a gene is duplicated by any mechanism, the expected consequence is that one or the other of the duplicates will be lost. However, too many duplicates somehow resist loss [14–17]. The mutational process that leads to this sort of gene loss is called 'fractionation', or sometimes 'diploidization' when the duplication is by tetraploidy. Fractionation is mutation that removes parts or all of a duplicate gene's function. The relative contribution of point mutation versus deletion to the molecular-level fractionation mechanism is not yet known

[18], but many studies in higher plants have found that chromosomal aberrations are elevated following artificial or recent tetraploidy, especially allotetraploidy [19–23]. As with transcriptional dysfunction in tetraploids [24], these chromosomal dysfunctions may result from extreme heterozygosity (from a wide cross) and not tetraploidy per se [19, 25].

Among fully sequenced higher eukaryotic genomes, a random gene has a 10–30% chance of being in a local array, usually tandem, and a higher chance if alignment criteria are relaxed. Deducing duplication rates from frequency data without data from out-group genomes is confounded by gene conversion [26], so the exact meaning of tandem duplication frequencies is unknown. On the contrary, all genes in the genome that were duplicated by tetraploidy (homeologs) originated contemporaneously as pairs. There are a few examples of tetraploidies that occurred long-enough ago that the genomes have fractionated back to near-diploidy, but are not so ancient as to obscure history. The probable pair of tetraploidies at the base of the chordate lineage (2R) exemplifies such ancient, obscure (in the 'twilight zone') events [27, 28]. More recent tetraploidies occurred in the *Xenopus laevis* lineage, leaving less than 10% of the genome in pairs [29], and in the teleost fish lineage leaving 20–24% in pairs when measured in *Tetraodon* [30]. The fish tetraploidy includes both lineages of teleosts, placing it approximately 300 million years ago [31]. Tetraploidy has been commonplace in the flowering plant lineage, at least for the last 200 MY. Approximately every 60 million years a tetraploidy occurred in typical plant lineages, and very recent, unfractionated tetraploidies are commonplace in plants. The flowering plant *Arabidopsis*, like every other dicot or monocot, contains syntenic blocks evidencing multiple large-scale duplication events. According to Bowers and coworkers [32] – independently supported by our own work [18] – the most recent tetraploidy, called alpha (α), now covers about 85% of the genome; 24.5% of the current genome is composed of pairs of genes retained from this event. (Thus, *Arabidopsis* α-retention frequency is said to be 24.5%). While the poplar genome carries two tetraploidies in common with the *Arabidopsis* genome, its most recent tetraploidy is in its own lineage; at least 35% of the poplar genome is retained [33]. The rice genome carries at least one ancient tetraploidy in common with *Arabidopsis*, poplar and presumably all flowering plants [32, 34]. The most recent tetraploidy in the rice lineage happened just before the grass radiation; 22–50% of the rice genome is paired today [35–37]. In summary, between 10 and 50% of 'fully' fractionated post-tetraploid eukaryotic gene content, in post-tetraploid lineages, exists as pairs derived from the most recent tetraploidy.

A third type of duplication involves chromosomal segments. Although important in all lineages, segmental duplications should be especially important in those many animal lineages without tetraploidy. For example, many segmental duplications characterize one difference between the genomes of chimpanzee and man [38]. An analysis of changes in gene content following segmental duplications is beyond the scope of this critique.

On the Mechanism of Duplicate Gene Retention

Given these significant levels of retained duplications (both homeologous/post-tetraploidy and local), one or more retention mechanisms must operate, as presented previously. The classic scheme where selection is relaxed and one of the duplicates sometimes mutates to a new, adaptive function – the Gain-of-Function scheme – is too slow to avoid loss by purifying selection [16]. Subfunctionalization, on the other hand, may also be too slow unless population sizes are small [39], but the two mutational hits required for subfunctionalization to work result in selectively neutral losses-of-function. These would be the sort of knockouts expected if fractionation involves either point mutation or insertion/deletion [14]. The beauty of subfunctionalization is that a pair of genes can become 'locked-in' without positive selection for any new function. According to the ISI Web of Knowledge citation index, the Force and coworkers study [2] had been cited 698 times between 1999 and October, 2006. Subfunctionalization is the current, accepted explanation for duplicate gene retention.

Dozens of large-scale experiments provide overwhelming evidence that subfunctionalization can and often does occur, and has been reviewed [6, 40, 41]. These data involve subfunctionalization as a mechanism, not Subfunctionalization as a hypothesis explaining retention (and capitalized for that reason). This same literature documents the many cases that measure one or both genes evolving faster post-duplication than expected had they been singlets. For *Arabidopsis* homeologs, an interesting hypothesis posits that pairs regulating development diverge slowly while pairs responding to environmental signals diverge rapidly [42]. Neither asymmetry of base substitution rate nor mutation rate acceleration is universal when comparing duplicate gene divergence [43, 44]. Certainly, when duplicates diverge, and especially when they diverge by subfunctionalization, retention becomes more durable. It also seems plausible that duplicates retained for *any* reason will diverge to useful function, and resist subsequent loss. The question is: does the subfunctionalization mechanism generally retain genes that would otherwise be lost after tetraploidy? Or, does subfunctionalization operate largely *after* genes are retained, thereby playing a lesser role in changes of genc content? Certainly, subfunctionalization tends to happen once genes are retained. To confuse retention with post-retention issues commits the rhetorical fallacy of *Ignoratio elenchi*.

Published Data On How Duplications Change Gene Content in Eukaryotes

The mechanism of gene duplicate retention is particularly important for evolution because this mechanism is positioned so as to have the potential to change gene content, not just deepen the gene pool with new alleles of the same genes. In 1999, the year of the origination of the Subfunctionalization hypothesis, there were almost no gene-content data involving duplications. In any case, there was not – and still is not – an easy way to predict what types of gene would be more 'subfunctionalizable' than any other. There are now data on the category of gene – usually estimated by Gene Ontology (GO) annotation – that tend significantly to be duplicated locally (in tandem) and/or that tend significantly to be retained following tetraploidy. Among the various web applications that help the researcher evaluate statistically whether or not any gene list carries over- or under-represented GO terms – with some control over the problem of nesting – GOstat [45] is particularly useful. The use of GO terms is often frustrating because a high proportion of the genes in a list are vaguely or not annotated, functionally specific GO categories often have too few genes in them to support statistical analysis and because relatively recent (<100 MYA) homeologous pairs in plants often carry different GO annotations (which constitutes error). Given this caveat, I will review published evidence and provide some new evidence that fractionation of any sort of gene/genome duplication is biased, leading to expansion of some gene families but not others. The data for *Arabidopsis* and rice tetraploidies, and tandem duplications in *Arabidopsis* and a few vertebrates was recently evaluated and documented [review 6]. In general, genes that retain their tandem duplicates are genes whose products have few interactions with other genes' products or play metabolic or specific degradation-repair roles. Genes significantly *not* retained in tandem are regulatory, developmental or genes whose products have many interactions. Recently [46] the Gaut laboratory compared the gene contents of local duplicates derived from the monocot rice and the dicot *Arabidopsis*. They found striking similarities involving enrichment for 'membrane proteins' and genes encoding proteins induced by 'abiotic and biotic stress', and also found similarities of under-representation for genes involved in 'transcription'. This is in general agreement with previous results on plants and animals [6]. Genes whose products are transcription factors tend to *not* be retained in tandem. In all cases measured, retention following local duplication is significantly biased as to gene function.

As indicated, there are now three plant and two animal higher eukaryotic genomes that can be analyzed for gene content changes following an ancient (but still tractable) tetraploidy. The *Arabidopsis* tetraploidy is best understood.

In general, and as predicted by the Hurst lab [9] from work in yeast, by the J.A. Birchler lab from work on particular genes in maize and *Drosophila* [8] and from R.A. Veitia's [10, 47] more theoretical work on transcription factors in mammals, genes that are 'connected' such as within a protein complex, or in a regulatory loop or found participating in any 'web of dependency', are over-retained, and genes whose products act alone tend to be under-represented among retained pairs [6, 34, 48, 49]. Going further, my group [18] found that retained genes in *Arabidopsis* were clustered on chromosomes. Clusters occurred because one homeolog lost more genes than the other, forming clusters naturally on the over-fractionated chromosome. Those genes most clustered were genes with 'ubiquitin' somewhere in their GO description, the proteasome being a particularly complex machine, and genes encoding transcription factors. The poplar genome has no GO annotations. Deductions from *Arabidopsis* best blast hits found that the pairs retained post-tetraploidy are similar to those of *Arabidopsis*, except that 'structural component of the ribosome' was not over-represented (Hellsten, Pedersen, Rokhsar and Freeling, unpublished). Rice 'transcription factors' are vastly over-retained post tetraploidy [36], just as they are in *Arabidopsis* and poplar. Comparisons among retentions following the three sequential tetraploidies in the *Arabidopsis* lineage [32] led Seoighe and Wolfe [49] to the conclusion that genes retained at one tetraploidy would tend to be retained once again at the next.

The teleost fish *Tetraodon* tetraploidy retention dataset has very few retained genes in it [30]. For example, although the GOSlim biological process term 'communication' is significantly (1.8-fold, $p = 4e^{-6}$) over-retained, the total retained gene count is 43. The molecular function GO term 'nucleic acid binding' is *under*-retained (1.54-fold; $p = 0.02$) but their retained gene list includes only 13 of these genes. This same GO term is significantly over-retained post tetraploidy in plants. Although there must be thousands of genes encoding transcription factors in the fish, this particularly specific GO term was not included on the list. In general, there are too few genes from the fish tetraploidy to be useful. The frog lineage tetraploidy [29] is far more recent than that in the fish. The retained gene list is also short. These workers found almost no indications of gene content bias post-tetraploidy, and presented no evidence that 'transcription factor activity' is over-retained. An independent study of 290 post-tetraploidy pairs in *Xenopus laevis* (the frog tetraploid) [50] found cases of subfunctionalization and neofunctionalization, but the retained genes themselves were not significantly different from the presumed ancestor. These frog data are unique in that they evidence no statistically significant post-tetraploidy gene retention and they contain very small retained gene sets. The frog data neither support nor refute Subfunctionalization as an important mechanism explaining duplicate gene retention. Balanced Gene Drive does not explain the frog data, but it is the best among the three mechanisms under consideration as the primary retention mechanism.

Two New Case Studies: Tests of the Reciprocity of Gene Content Changes Following Tandem versus Whole Genome Duplication in *Arabidopsis*

Neither Gain-of-Function nor Subfunctionalization predicts that there will be any sort of gene, any GO category, preferentially retained post duplication, and – even if there were – there should be a *positive* correlation between retention post local duplication and retention post-tetraploidy. Alternatively, Balanced Gene Drive predicts that dose-sensitive genes should tend to not be duplicated locally, but – reciprocally – tend to be retained post-tetraploidy. We [51] have published an *Arabidopsis* gene list annotating each gene with our estimate as whether or not it is a local duplicate (using, in part, data from [52]), whether or not it has an α-pair retained from the most recent tetraploidy, and the DATF [53] transcription factor subfamily to which it belongs. GO terms were from TAIR, as incorporated into the GOstat statistical application (GOstat settings: $p < 0.001$; clusters = –1 meaning that we did not identify nests of terms). Figure 1 shows all molecular function GO terms from *Arabidopsis* that were significantly either over- (O) or under-retained (U) in either the tandem (local) duplicate list (T for tandem: the leftmost column of data in fig. 1) or the α-duplicate gene list (A for 'alpha'). The positive correlation of gene content change between tandem and tetraploidy duplication, highlighted medium grey, is not the general rule. However, neither is the reciprocal relationship (dark grey) the general rule, although genes encoding transcription factors and ribosomal proteins do exhibit reciprocity. The general rule is that a GO category that is significantly over- or under-represented after either tandem duplication or tetraploidy is not significantly changed (N) in gene content in the comparator gene list; GO terms associated with genes exhibiting this 'lack of relationship' general rule are coded light grey. Note that several dramatic reciprocal frequencies are judged to be 'not significant' by GOstat because there are too few genes in either the tandem (T list) or alpha (A list) categories (fig. 1).

There are many problems with the GO method of functional annotation. The GO terms listed in figure 1 have overlapping gene contents. In fact, a clustering of related molecular function terms that show a reciprocal relationship are actually those with transcription factor genes and ribosomal protein genes only. Similarly, those GO terms that are positively correlated (medium grey) are actually only three sorts of genes: kinases (not specifically protein-kinases), pectinesterases and cation-transporters.

List A=alpha duplicates, T=local duplicates	GO Term	Number of genes/genome recognized by GOStat	number of locally repeated genes (>5 if O)	p (local duplicates); Significance Cutoff=0.05=5E-2	Significant O, U or N in Tandem	frequency of retention as a local duplicate	Molecular Function GO Description (nests ignored)	Significant O or U in alpha homeologs? N=not significant	Number of genes in alpha pairs	p (alpha tetraploidy)	frequency of retention as alpha pair
A	GO:0005554	11632	810	4.90E-01	N	0.07	Molecular function unknown	U	1472	1.49E-04	0.13
T	GO:0008324	429	10	3.60E-03	U	0.02	Cation transporter activity	U	35	1.64E-02	0.08
A+T	GO:0004871	331	8	1.75E-02	U	0.02	Signal tranducer activity	U	21	2.90E-02	0.06
A	GO:0008514	21	0	1.00E+00	N	0.00	Organic anion transporter activity	O	15	1.46E-07	0.71
A	GO:0004723	9	1	6.69E-01	N	0.11	Calcium-dependent serine-threonine phosphatase activity	O	6	9.60E-03	0.67
A+T	GO:0046910	64	16	6.22E-05	O	0.25	Pectinesterase inhibitor activity	O	22	1.70E-04	0.34
A	GO:0030599	179	11	5.90E-02	N	0.06	Pectinesterase activity	O	21	6.05E-03	0.12
A	GO:0008509	107	3	3.58E-01	N	0.03	Anion transporter activity	O	31	3.72E-04	0.29
A	GO:0005525	179	5	1.88E-01	N	0.03	GTP binding	O	44	1.44E-03	0.25
A	GO:0046914	699	36	2.88E-01	N	0.05	Transition metalion binding	O	168	5.33E-13	0.24
A	GO:0008270	609	30	2.56E-01	N	0.05	Zinc ion binding	O	145	6.87E-11	0.24
T	GO:0003735	427	4	3.94E-05	U	0.01	Structural constituant of ribosome	O	101	2.53E-07	0.24
A	GO:0043169	875	46	2.60E-01	N	0.05	Cation binding	O	206	9.09E-15	0.24
A	GO:0046872	940	51	3.13E-01	N	0.05	Metal ion binding	O	211	1.06E-12	0.22
A	GO:0043167	940	51	3.13E-01	N	0.05	Ion binding	O	211	1.06E-12	0.22
A+T	GO:0005198	536	7	1.14E-05	U	0.01	Structural molecule activity	O	118	2.04E-06	0.22
A+T	GO:0003700	2056	72	4.55E-08	U	0.04	Transcription Factor activity	O	423	4.45E-17	0.21
A+T	GO:0016301	1608	163	8.22E-07	O	0.10	Kinase activity	O	328	8.41E-13	0.20
A+T	GO:0030528	2309	73	6.12E-11	U	0.03	Transcription Regulator activity	O	449	1.20E-13	0.19
A+T	GO:0016772	1823	199	2.46E-11	O	0.11	Transferase activity; trans. P-containing groups	O	353	1.92E-10	0.19
A+T	GO:0003677	2675	107	1.03E-07	U	0.04	DNA binding	O	517	3.20E-15	0.19
A	GO:0016773	702	39	4.60E-01	N	0.06	Phosphotransferase activity, alcohol group as acceptor	O	131	6.05E-03	0.19
A	GO:0005515	1474	76	8.46E-02	N	0.05	Protein binding	O	265	1.35E-04	0.18
T	GO:0003676	3545	132	3.16E-12	U	0.04	Nucleic acid Binding	O	618	4.70E-09	0.17
A+T	GO:0005488	6789	391	3.35E-03	U	0.06	Binding	O	1164	5.46E-16	0.17
A+T	GO:0016740	3042	291	3.60E-09	O	0.10	Transferase activity	N	491	4.00E+00	0.16
A	GO:0003674	28874	1981	2.20E-01	N	0.07	Molecular function	O	4100	8.15E-04	0.14
T	GO:0004126	9	6	1.02E-04	O	0.67	Cytidine deaminase activity	N	2	6.20E-01	0.22
T	GO:0030234	236	30	4.03E-03	O	0.13	Enzyme regulator activity	N	51	1.64E-02	0.22
T	GO:0004650	77	15	3.48E-04	O	0.19	Polygalacturanidase activity	N	16	3.40E-01	0.21
T	GO:0004857	118	28	3.25E-11	O	0.24	Enzyme inhibitor activity	N	24	2.60E-01	0.20
T	GO:0008194	195	35	2.70E-08	O	0.18	UDP-glycosyl transferase activity	N	36	3.00E-01	0.18
T	GO:0004721	236	5	4.54E-02	U	0.02	Phosphoprotein phosphatase activity	N	41	3.00E-01	0.17
T	GO:0016791	362	6	2.02E-03	U	0.02	Phosphoric monoester hydrolse activity	N	60	4.00E-01	0.17
T	GO:0016789	330	37	1.42E-02	O	0.11	Carboxylic ester hydrolase activity	N	54	5.00E-01	0.16
T	GO:0000166	887	37	1.75E-02	U	0.04	Nucleotide binding	N	144	2.50E-01	0.16
T	GO:0004175	222	29	3.34E-03	O	0.13	Endopeptidase activity	N	36	6.30E-01	0.16
T	GO:0005516	212	4	4.76E-02	U	0.02	Calmodulin binding	N	34	7.00E-01	0.16
T	GO:0004185	69	11	4.39E-02	O	0.16	Serine carboxypeptidse activity	N	11	9.00E-01	0.16
T	GO:0042578	408	9	3.61E-03	U	0.02	Phosphoric ester hydrolase activity	N	65	5.20E-01	0.16
T	GO:0016787	2744	233	2.14E-03	O	0.08	Hydrolse activity	N	412	2.90E-01	0.15
T	GO:0005215	1504	64	1.00E-03	U	0.04	Transporter activity	N	223	5.00E-01	0.15
T	GO:0003824	8333	764	6.06E-22	O	0.09	Catalytic activity	N	1216	1.79E-01	0.15
T	GO:0016798	476	69	8.31E-10	O	0.14	Hydrolase activity, acting on glycosyl bonds	N	68	9.92E-01	0.14
T	GO:0019239	14	6	2.14E-03	O	0.43	Deaminase activity	N	2	1.00E+00	0.14
T	GO:0008236	182	32	2.76E-07	O	0.18	Serine-type peptidase activity	N	25	1.00E+00	0.14
T	GO:0004553	421	63	9.22E-10	O	0.15	Hydrolase activity, hydrolyzing o-glycosyl compounds	N	57	1.00E+00	0.14
T	GO:0015075	594	16	1.00E-03	U	0.03	Ion transporter activity	N	71	5.00E-01	0.12
T	GO:0004289	62	18	1.63E-06	O	0.29	Subtilase activity	N	7	9.00E-01	0.11
T	GO:0016779	198	36	9.24E-09	O	0.18	Nucleotidyltransferase activity	N	21	5.00E-01	0.11
T	GO:0004252	96	20	2.11E-06	O	0.21	Serine-type endopeptidase activity	N	10	6.90E-01	0.10
T	GO:0003723	487	15	1.41E-02	U	0.03	RNA binding	N	49	3.20E-01	0.10
T	GO:0016814	21	6	1.70E-02	O	0.29	Hydrolase act., C-N bonds, not peptide, in clyclic amidines	N	2	9.00E-01	0.10
T	GO:0030246	72	26	2.46E-11	O	0.36	Carbohydrate binding	N	6	5.00E-01	0.08
T	GO:0019825	253	77	3.17E-48	O	0.30	Oxygen binding	N	21	1.00E-01	0.08
T	GO:0004866	40	9	1.00E-02	O	0.23	Endopeptidase inhibitor activity	N	2	4.00E-01	0.05
T	GO:0030414	45	12	4.39E-04	O	0.27	Protease inhibitor activity	N	2	3.70E-01	0.04

Significant concordant gene content (O-O or U-U) between local and post-tetraploid duplications. **O= Over- U=Under-representation.**
Significant reciprocal gene content (O-U or U-O) between local duplications and post-tetraploid duplications
No significant relationship of gene content (0-N, U-N), as if independent mechanisms retained local vs. post-tetraploid duplications. **N= Not sig.**

Fig. 1. The relationship between GO categories of *Arabidopsis* genes that were significantly over- (O) or under- (U) retained in after either local (tandem) duplication (List T) or after the most recent tetraploidy (List A). The grey tone code – with legend embedded in the Figure – indicates the nature of the relationship between retentions in Lists T and A, and their significance. Statistics by GOstat (see text).

Transcription factors are important for regulatory evolution, and – if there is one result about duplication that is singularly descriptive – transcription factor genes are over-retained post-tetraploidy and under-retained post-tandem duplication (at least in plants). However, anyone who has browsed through any eukaryotic genome knows that sometimes transcription factor genes are duplicated locally. The Database of *Arabidopsis* Transcription Factors (DATF) contained 1826 genes as of September 2005 (http://datf.cbi.pku.edu.cn/). DATF sorted each TF gene to one of 56 TF subfamilies. Our α-pair analysis [54] added 25 new genes for a total revised DATF gene count of 1851; see figure 2 legend (because we used TF post-tetraploidy retention data published previously, this experiment does not use the most recent edition of DATF). We confined our analysis to those 22 families with 23 genes or more in order to avoid scatter caused by small numbers. We then plotted frequency of retention following local duplication on the Y-axis. Units are defined exactly in the legend. Figure 2 shows a clear reciprocal relationship: transcription factor families that are relatively well retained following tetraploidy are poorly retained following tandem duplication. Using the R statistical package, the regression line of figure 2 has a 0.01 probability of being slope 0. The DATF abbreviation for each transcription factor family, followed by the total number of genes in the genome, describe each point in figure 2.

Balanced Gene Drive, as defined previously, explains the reciprocal data of figure 2. Loss of dose-sensitive post-tetraploidy homeologs could have occurred if all of the inter-dependent genes had been simultaneously fractionated, but there is no mechanism known able to accomplish coordinated gene removals. These same dose-sensitive genes, when duplicated tandemly, immediately upset the balance of gene products, with selectively negative effects. Balanced Gene Drive predicts the reciprocal data given in figure 2. Subfunctionalization predicts a positive correlation for figure 2, and is clearly not the retention mechanism for plant transcription factor genes.

Continued research on the data of figure 2 might attempt to categorize TF genes not only by their DNA-binding motifs but by essential sort of downstream biology. Ha and coworkers [42] inferred that homeologous genes in *Arabidopsis* involved in development were less subfunctionalized than those pairs involved in 'response to …' various exogenous signals. For example, MADS-box TFs seem to be 'developmental' and HB-Zip I and II genes seem to be 'response to ...'. It would be especially interesting to know whether or not 'subfunctionalizability' of homeologs extended to subfunctionalizability of tandem duplicates. Once retained by any mechanism, redundant genes that are particularly subfunctionalizable are expected to subfunctionalize, so the results of this continued research are not likely to reinstall subfunctionalization as *the* mechanism of retention.

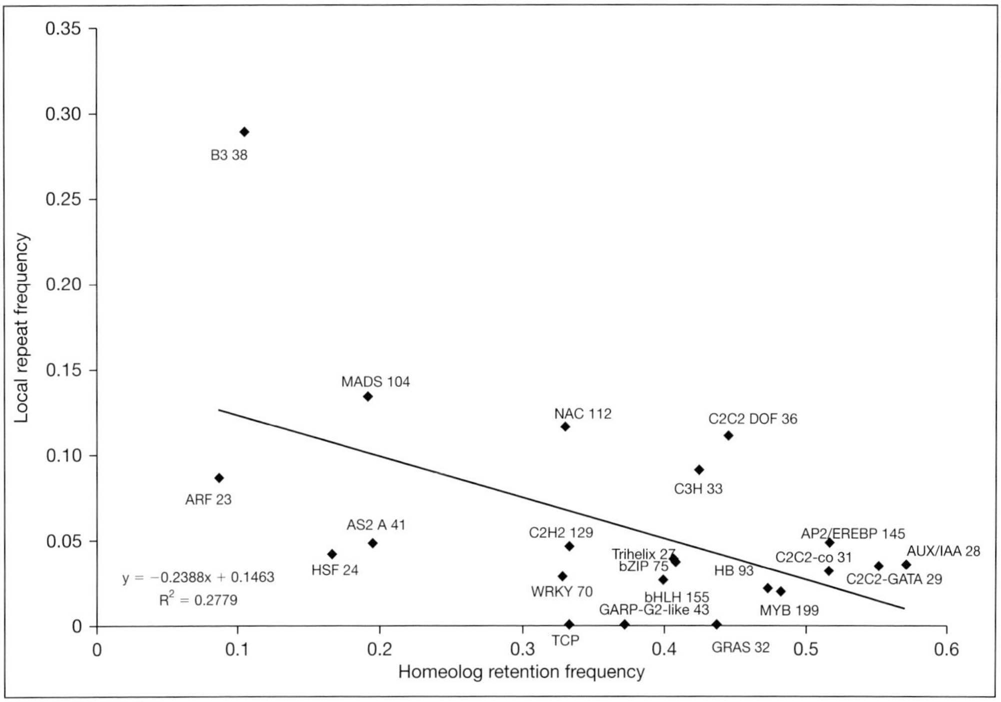

Fig. 2. The reciprocal relationship between duplicate retention frequencies following local (tandem) duplication (Y-axis) versus the retention frequency of these same genes post-α-tetraploidy for *Arabidopsis* transcription factor family genes. Gene list and retention data from Supplementary Information 1 of one of my own publications [51]. This research used families of transcription factors from the Database of *Arabidopsis* Transcription Factors (DATF), July 2005. The organization and total numbers of DATF transcription factors have now been revised, but not dramatically. Families with total gene numbers ≥23 were used; this number decorates each point on the graph following the DATF family abbreviation. A particularly revised DATF subfamily is the 2005 B3 family. In the current DATF edition (not used), B3 family is subsumed within a revised AB13-VP1 family now containing 59 genes; the new datapoint – using updated DATF data – would be Y axis = 0.24 and the X axis = 0.15 yielding a point in approximately the same place as the B3 point in this figure. Unit Y axis is number of duplicates/total number of genes. Unit X axis is genes in alpha pairs/total number of genes. There are few local duplications and about 80% of them are tandem doublets; removing the duplicates from the total number of genes 'available' for retention post-tetraploidy does not significantly alter the regression line. Some duplicate genes are also retained as alpha pairs.

There Are No Data in Support of Subfunctionalization as A Duplication Retention Mechanism. What about the Duplication Data from the Frog, and other Data That Are Inconsistent with the Balanced Gene Drive Mechanism?

Five molecular function *Arabidopsis* GO categories, two clusters, are consistent with the Subfunctionalization retention mechanism. These terms are the medium grey rows in figure 1, indicating a positive relationship between retention and type of duplication. The general result is 'no relationship,' which does *not* support Subfunctionalization. If a gene is 'subfunctionalizable,' it makes sense that it would be so following any sort of duplication. There is no empirical evidence for a Subfunctionalization explanation for post-duplication retention. For example, the more chromosome space a gene uses in its *cis*-function, the more 'subfunctionalizable' it may be, whether the duplicates originated tandemly or post-tetraploidy. This hypothesis is now testable in *Arabidopsis* since the length of many genes can be estimated using the distribution of intragenic conserved noncoding sequences; genes in some *Arabidopsis* gene families stand alone surrounded by many kbs of functional chromosome while most *Arabidopsis* genes are compact and positioned near other genes [51, 54].

The reciprocal relationship between retention frequencies when tandem duplications are compared with post-tetraploidy duplications supports the Balanced Gene Drive explanation specifically for two *Arabidopsis* GO clusters (fig. 1) and among *Arabidopsis* transcription factor families in general (fig. 2). However, neither Balanced Gene Drive, nor any other mechanism discussed so far, accounts for most of the significant under- and over-retained GO terms (defined by gene lists) of figure 1 or in the literature in general.

The post-tetraploid retained gene datasets for frog and fish are just too small to justify a serious animal-plant comparison. Higher plants and animals respond to gene dosage very differently, so there is really no firm expectation that dosage will prove to be as important for animal gene duplication retentions as it has done in plants; the animal soma is more reactive to gene dosage imbalance than is the plant soma, but the plant gametophyte is far more sensitive to gene dosage than is the inert animal gamete. Animal genomes that went tetraploid within a window useful for analysis – approximately 120–40 MYA – are particularly important for continued and further analysis.

How come genes encoding 'structural constituents of the ribosome' – the first dark grey (reciprocal) row of figure 1 – is significantly over-retained in *Arabidopsis* and rice, but not in poplar or, apparently, either vertebrate? Papp and coworkers [9], in their yeast haploinsufficiency article, showed that yeast genes encoding ribosomal proteins were over-represented among retained pairs from the yeast tetraploidy. The reason for dose-sensitivity was that the exact concentrations of protein mattered because they had to make a complex with many partners in a strict stoichiometry, thereby creating a 'web of dependency'. This is a prediction of the Gene Balance Hypothesis, which has been reviewed [6, 8]. So, how come all genomes do not over-retain genes encoding ribosomal subunits post-tetraploidies? The answer must be that some genomes have evolved ways of mitigating the dosage effects that naturally accompany particular protein-protein complexes. Veitia [47] explores a number of mechanisms that could mitigate dosage effects. For example, if a chaperon concentration limits the assembly of a complex where protein subunits were limiting previously, a dosage effect is mitigated. There is much need for research on how dosage effect phenotypes have been mitigated since a retained pair that contributes to a gain-of-function must, of necessity, have somehow broken the web-of-dependency in order to have evolved.

Mechanisms other than avoidance of dosage effects could retain a nonrandom set of genes post-duplication, and thus contribute to gene family expansions. In yeast, there is a tendency for highly-expressed genes to be retained after the most recent tetraploidy [55], and it is plausible that particular genes will prove to be prone to silencing/epigenetic marking [18] or display particularly large targets for fractionation [51]. Removing Subfunctionalization from the duplication-retention position makes room for mechanisms that could – until mitigated – move the trajectory of evolution in one direction or another. Any research explaining trends or major adaptations seems to me to be of the highest priority. This way of thinking was intentionally banned from consensus evolutionary theory [11] along with the radical (but still fresh) Mutationist ideas of R.B. Goldschmidt [56]. That was about 60 years ago, before evolutionary theory was challenged to explain comparative gene content data.

A Place for Subfunctionalization in the Evolution of Novelty

Of the three mechanisms for post-duplication retention being considered here, only Balanced Gene Drive makes significant progress toward explaining changes in gene content following duplications in eukaryotic genomes. Balanced Gene Drive is particularly successful in predicting data for genes encoding transcription factor families and ribosomal proteins. Even so, continued research will probably identify other mechanisms, perhaps epigenetic mechanisms, that further account for trends in gene family expansion following gene duplication. Moreover, neither Balanced Gene Drive nor Subfunctionalization generate any new functions, only multiple genes. Gain-of-Function (neofunctionalization) directly underlies the origin of novelty. Clearly, Gain-of-Function is a multi-step process. In some cases, Gain-of-Function requires duplicate and diverged modules of functionally-interacting genes; just one gene doing something new would be inadequate. Subfunctionalization can play an important role as a 'division of labor' step in this process. This important role in the evolution of novelty has been discussed in several recent publications [41, 57–59]. Although downgraded from a gene retention mechanism to a gene divergence mechanism primarily, subfunctionalization remains an important mechanism contributing to duplicate gene evolution.

Acknowledgement

Thanks to our research team and particularly collaborator Brian C. Thomas and student Eric Lyons for critically reading this work. I thank Damon Lisch for discussions. This research was funded by US NSF grant DBI-034937.

References

1 Lespinet O, Wolf YI, Koonin EV, Aravind L: The role of lineage-specific gene family expansion in the evolution of eukaryotes. Genome Res 2002;12:1048–1059.
2 Force A, Lynch M, Pickett FB, Amores A, Yan YL, Postlethwait J: Preservation of duplicate genes by complementary, degenerative mutations. Genetics 1999;151:1531–1545.
3 Hughes AL: The evolution of functionally novel proteins after gene duplication. Proc Biol Sci 1994;256:119–124.
4 Lewis EB: Pseudoallelism and gene evolution. Cold Spring Harb Symp on Quant Biol 1951;16:159–174.
5 Ohno S: Evolution by Gene Duplication. Berlin, Springer-Verlag, 1970.
6 Freeling M, Thomas BC: Gene-balanced duplications, like tetraploidy, provide predictable drive to increase morphological complexity. Genome Res 2006;16:805–814.
7 Birchler JA, Auger DL, Riddle NC: In search of the molecular basis of heterosis. Plant Cell 2003;15:2236–2239.
8 Birchler JA, Riddle NC, Auger DL, Veitia RA: Dosage balance in gene regulation: biological implications. Trends Genet 2005;21:219–226.
9 Papp B, Pal C, Hurst LD: Dosage sensitivity and the evolution of gene families in yeast. Nature 2003;424:194–197.
10 Veitia RA: Exploring the etiology of haploinsufficiency. Bioessays 2002;24:175–184.
11 Mayr E: What was the Evolutionary Synthesis? Trends Ecol Evol 1993;8:31–34.
12 McShea DW: Possible largest-scale trends in organismal evolution: Eight 'live hypotheses'. Annu Rev Ecol Syst 1998;29:293–310.
13 Maynard Smith J, Szanthmary E: The Major Transitions in Evolution. Oxford, Freeman, 1995.
14 Lynch M: Genomics. Gene duplication and evolution. Science 2002;297:945–947.
15 Lynch M, Conery JS: The evolutionary fate and consequences of duplicate genes. Science 2000;290:1151–1155.
16 Lynch M, Force A: The probability of duplicate gene preservation by subfunctionalization. Genetics 2000;154:459–473.
17 Fischer R: The sheltering of lethals. Am Nat 1935;69:446–455.
18 Thomas BC, Pedersen B, Freeling M: Following tetraploidy in an *Arabidopsis* ancestor, genes were removed preferentially from one homeolog leaving clusters enriched in dose-sensitive genes. Genome Res 2006;16:934–946.
19 Hegarty M, Barker G, Wilson I, Abbott R, Edwards S: Transcriptome shock after interspecific hybridization in *Senecio* is ameliorated by genome duplication. Curr Biol 2006;16:1652–1659.

20 Shaked H, Kashkush K, Ozkan H, Feldman M, Levy AA: Sequence elimination and cytosine methylation are rapid and reproducible responses of the genome to wide hybridization and allopolyploidy in wheat. Plant Cell 2001;13:1749–1759.

21 Kashkush K, Feldman M, Levy AA: Gene loss, silencing and activation in a newly synthesized wheat allotetraploid. Genetics 2002;160:1651–1659.

22 Osborn TC, Pires JC, Birchler JA, Auger DL, Chen ZJ, et al: Understanding mechanisms of novel gene expression in polyploids. Trends Genet 2003;19:141–147.

23 Song K, Lu P, Tang K, Osborn TC: Rapid genome change in synthetic polyploids of *Brassica* and its implications for polyploid evolution. Proc Natl Acad Sci USA 1995;92:7719–7723.

24 Chen ZJ, Ni Z: Mechanisms of genomic rearrangements and gene expression changes in plant polyploids. Bioessays 2006;28:240–252.

25 Wang J, Tian L, Lee HS, Wei NE, Jiang H, et al: Genomewide nonadditive gene regulation in *Arabidopsis* allotetraploids. Genetics 2006;172:507–517.

26 Gao LZ, Innan H: Very low gene duplication rate in the yeast genome. Science 2004;306:1367–1370.

27 Simillion C, Vandepoele K, Saeys Y, Van de Peer Y: Building genomic profiles for uncovering segmental homology in the twilight zone. Genome Res 2004;14:1095–1106.

28 McLysaght A, Hokamp K, Wolfe KH: Extensive genomic duplication during early chordate evolution. Nat Genet 2002;31:200–204.

29 Morin RD, Chang E, Petrescu A, Liao N, Griffith M, et al: Sequencing and analysis of 10,967 full-length cDNA clones from *Xenopus laevis* and *Xenopus tropicalis* reveals post-tetraploididation transcriptome remodeling. Genome Res 2006;16:796–803.

30 Brunet FG, Crollius HR, Paris M, Aury JM, Gibert P, et al: Gene loss and evolutionary rates following whole-genome duplication in teleost fishes. Mol Biol Evol 2006;23:1808–1816.

31 Taylor J, Braasch I, Frickey T, Meyer A, Van de Peer Y: Genome duplication, a trait shared by 22,000 species of ray-finned fish. Genome Res 2003;13:382–390.

32 Bowers JE, Chapman BA, Rong J, Paterson AH: Unravelling angiosperm genome evolution by phylogenetic analysis of chromosomal duplication events. Nature 2003;422:433–438.

33 Tuskan GA, Difazio S, Jansson S, Bohlmann J, Grigoriev I, et al: The genome of black cottonwood, *Populus trichocarpa* (Torr. & Gray). Science 2006;313:1596–1604.

34 Maere S, De Bodt S, Raes J, Casneuf T, Van Montagu M, Kuiper M, Van de Peer Y: Modeling gene and genome duplications in eukaryotes. Proc Natl Acad Sci USA 2005;102:5454–5459.

35 Paterson AH, Bowers JE, Chapman BA: Ancient polyploidization predating divergence of the cereals, and its consequences for comparative genomics. Proc Natl Acad Sci USA 2004;101: 9903–9908.

36 Tian CG, Xiong YQ, Liu TY, Sun SH, Chan LG, Chen MS: Evidence for ancient whole genome duplication event in rice and other cerials. Yi Chuan Xue Bao 2005;32:519–527.

37 Yu J, Wang J, Lin W, Li S, Li H, et al: The genomes of *Oryza sativa*: a history of duplications. PLoS Biol 2005;3:e38.

38 Cheng Z, Ventura M, She X, Khaitovich P, Graves T, et al: A genome-wide comparison of recent chimpanzee and human segmental duplications. Nature 2005;437:88–93.

39 Ward R, Durrett R: Subfunctionalization: How often does it occur? How long does it take? Theor Popul Biol 2004;66:93–100.

40 Force A, Cresko WA, Pickett FB, Proulx SR, Amemiya C, Lynch M: The origin of subfunctions and modular gene regulation. Genetics 2005;170:433–446.

41 He X, Zhang J: Rapid subfunctionalization accompanied by prolonged and substantial neofunctionalization in duplicate gene evolution. Genetics 2005;169:1157–1164.

42 Ha M, Li WH, Chen ZJ: External factors accelerate expression divergence between duplicate genes. Trends Genet 2007;23:162–166.

43 Jordan IK, Wolf YI, Koonin EV: Duplicated genes evolve slower than singletons despite the initial rate increase. BMC Evol Biol 2004;4:22.

44 Chapman BA, Bowers JE, Feltus FA, Paterson AH: Buffering of crucial functions by paleologous duplicated genes may contribute cyclicality to angiosperm genome duplication. Proc Natl Acad Sci USA 2006;103:2730–2735.

45 Beissbarth T, Speed T: GOstat: Find statistically over-represented gene ontologies within groups of genes. Bioinformatics 2004;1:1–2.

46 Rizzon C, Ponger L, Gaut BS: Striking similarities in the genomic distribution of tandemly arrayed genes in *Arabidopsis* and rice. PLoS Comput Biol 2006;2:e115.
47 Veitia RA: Gene dosage balance in cellular pathways: Implications for dominance and gene duplicability. Genetics 2004;104:569–574.
48 Blanc G, Wolfe KH: Functional divergence of duplicated genes formed by polyploidy during *Arabidopsis* evolution. Plant Cell 2004;16:1679–1691.
49 Seoighe C, Gehring C: Genome duplication led to highly selective expansion of the *Arabidopsis thaliana* proteome. Trends Genet 2004;20:461–464.
50 Chain FJ, Evans BJ: Multiple mechanisms promote the retained expression of gene duplicates in the tetraploid frog *Xenopus laevis*. PLoS Genet 2006;2:e56.
51 Freeling M, Rapaka L, Lyons E, Pedersen B, Thomas BC: G-boxes, Bigfoot genes and environmental response: characterization of the intragenomic conserved noncoding sequences of *Arabidopsis*. Plant Cell 2007;19:1441–1457.
52 Haberer G, Hindemitt T, Meyers BC, Mayer KF: Transcriptional similarities, dissimilarities, and conservation of cis-elements in duplicated genes of *Arabidopsis*. Plant Physiol 2004;136: 3009–3022.
53 Guo A, He K, Liu D, Bai S, Gu X, Wei L, Luo J: DATF: A database of *Arabidopsis* transcription factors. Bioinformatics 2005;21:2568–2569.
54 Thomas BC, Rapaka L, Lyons E, Pedersen B, Freeling M: Intragenomic conserved noncoding sequences in *Arabidopsis*. Proc Natl Acad Sci USA 2007;104:3348–3353.
55 Seoighe C, Wolfe KH: Yeast genome evolution in the post-genome era. Curr Opin Microbiol 1999;2:548–554.
56 Goldschmidt RB: Evolution as viewed by one geneticist. Am Scientist 1953;40:84–98.
57 Roth C, Rastogi S, Arvestad L, Dittmar K, Light S, Ekman D, Liberles DA: Evolution after gene duplication: models, mechanisms, sequences, systems, and organisms. J Exp Zoolog B Mol Dev Evol 2007;308:58–73.
58 Rastogi S, Liberles DA: Subfunctionalization of duplicated genes as a transition state to neofunctionalization. BMC Evol Biol 2005;5:28.
59 Prince VE, Pickett FB: Splitting pairs: the diverging fates of duplicated genes. Nat Rev Genet 2002;3:827–837.

Michael Freeling
Department of Plant and Microbial Biology, University of California
111 Koshland Hall
Berkeley, CA 94720 (USA)
Tel. +1 510 642 8058, Fax +1 510 642 4995, E-Mail freeling@nature.berkeley.edu

Volff J-N (ed): Plant Genomes.
Genome Dyn. Basel, Karger, 2008, vol 4, pp 41–56

Grass Genome Structure and Evolution

Joachim Messing[a], *Jeffrey L. Bennetzen*[b]

[a]Waksman Institute of Microbiology, Rutgers University, Piscataway, N.J.;
[b]Department of Genetics, University of Georgia, Athens, Ga., USA

Abstract

Advances in the construction of large insert libraries and physical maps have facilitated the sequencing of orthologous regions from related plant species that differ in genome size. This approach has been particularly productive for the Poaceae, including many important cereal crops. When the sequences of orthologous regions from closely related species are aligned, we can analyze the details of chromosomal evolution. The dynamics of chromosome structure appears to be driven by two types of rearrangement mechanisms, 'cut and paste' and 'copy and paste'. The latter mechanism has contributed to the expansion of orthologous regions, primarily by transposon amplification, while ongoing deletions by illegitimate and homologous recombination have at least partially counteracted or reversed this expansion in some regions. This review describes the current status of our understanding of the plasticity of plant genomes, emphasizing maize as a model for these studies.

The grass family of plants, the Poaceae, contains about 10,000 species, including some of the most important crops on earth, collectively referred to as the cereals. Grasses also exhibit some of the most intriguing features of chromosome evolution, including the C-value paradox, polyploidy, segmental duplication, gene amplification, and a level of dynamism unseen in other multicellular eukaryotes. Other families of plants have also undergone extensive chromosomal duplication and rearrangement [1–3], but the degree of variation found within the grasses, including haplotype variation in maize, stands out as exceptional [4]. This accelerated rate of change in genome structure after species divergence may help explain why the Poaceae have proven to be so adaptable to different environments, and certainly provides fertile ground for investigation of the nature and mechanisms of genomic instability.

Despite the great variation in genome size among the grasses, genetic maps based on conserved gene sequences indicate conservation of a rough collinearity

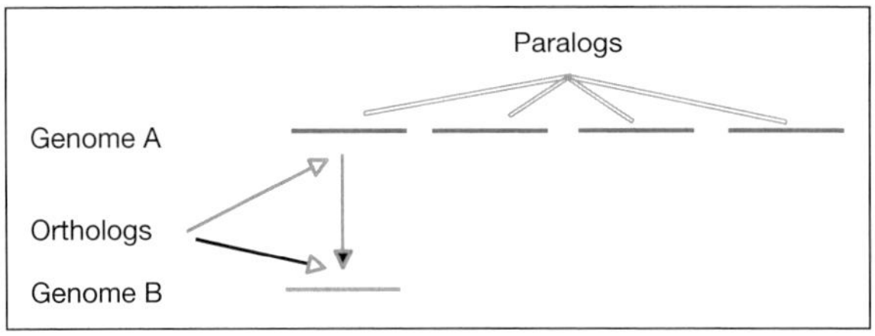

Fig. 1. Paralogous and orthologous DNA sequences. Explanation is given in the text.

of unique sequences within chromosomal segments [5]. Comparative genetic maps indicate that centromere position is often not conserved, suggesting a loss of centromeres and gain of neocentromeres thereby changing chromosome number and creating unique chromosomes composed of a mosaic of ancestral chromosomal segments for individual species. To reach a better understanding of these chromosomal rearrangements, it became critical to investigate gene collinearity in short contiguous DNA segments from the genome of a few representative species of the grass family. Perhaps the most unexpected result of these 'microcollinearity' studies was the discovery that many apparent genes have been recently mobile within cereal chromosomes.

What Is Orthologous and What Is Paralogous?

One complication to comparison of gene content and order in higher plants has been the very high frequency of tandem and non-tandem gene duplication. Hence, to determine whether gene order and content has been conserved in a given chromosomal segment, it is necessary to determine which gene family members were present in an ancestral species and which are more recent duplication products. Orthologous gene copies are those that are derived from ancestral loci by direct vertical descent. Hence, in comparing the genes in two separate species, all orthologous genes are expected to have begun to evolve independently at the same time, that being the point at which the two lineages that gave rise to these species first diverged from a shared ancestor. Paralogous gene copies are defined as those that have arisen by a copy and paste (i.e., duplicative) mechanism from orthologous genes (fig. 1). As a consequence, their degrees of sequence divergence should be less than those of orthologous ancestral genes, because their divergence could only begin after the time of gene duplication. For instance, many flowering plants have undergone whole-genome duplication events (WGDs), otherwise known as polyploidy, at some time in the evolution of their lineage. Therefore, their chromosomes initially contain duplicated regions with the same order of orthologous genes. Based on the mapping of unlinked gene duplicates, it was first recognized that maize was derived from a WGD [6, 7]. Genetic mapping of RFLP markers conserved in other grass species added support to these early observations [8, 9]. Comparative genomics of these markers showed that two maize chromosome sets aligned with one chromosome each of rice and sorghum, demonstrating that WGD truly did involve the whole genome [5, 10]. Furthermore, evolutionary analysis of duplicated genes also suggested that maize might have arisen by allotetraploidy, indicating the hybridization of two diverged progenitors rather than the duplication of a single progenitor [11]. Recent comparisons of orthologous chromosome intervals from these two chromosome sets with sorghum and rice showed that the two progenitors of maize and the progenitor of sorghum diverged at about the same time, ~11.9 million years ago (mya). This WGD in maize may have occurred as recently as 4.8 mya [12]. The implication is that gene copies in maize that diverged more recently than 4.8 mya are paralogous. Therefore, one could determine in any polyploidy species first the time of WGD before assessing whether a gene copy is likely to be orthologous or paralogous. Critical for this determination is that duplication of orthologous gene copies is representative for all chromosomes and has a common time point of divergence.

In maize, paralogs can be derived from either orthologous gene copy of the two progenitor chromosome sets. As an example, the *ZmCDA1* and *ZmCDA2* gene copies are orthologous [13] because they are descendants of the two progenitors of maize and their degree of nucleotide substitution is within the range which predicts that the two progenitors split about 11.9 mya [12]. Based on nucleotide variation, a paralogous gene copy, *ZmCDA3*, was derived from *ZmCDA1* about 4.5 mya. If this paralogous copy had arisen before speciation, then each copy would be expected on both maize homologs (neglecting subsequent deletion of one copy).

In a contrasting example, the *Fie* genes in maize were originally expected to be orthologous because they map as single copies to homoeologous chromosome locations. However, two tandem copies of *Fie* genes are observed in both sorghum and rice. Hence, the two progenitor chromosomal regions in maize had each preserved only one of the two tandem gene copies, and these were actually paralogs [12]. In this case, distinguishing between paralogs and orthologs was only accomplished by determining the degrees of nucleotide substitution for each copy of the tandem duplication in sorghum and rice. Otherwise, an error could have been made by assuming that the two maize *Fie* gene copies are orthologous because of their genetic map position.

A major challenge in the comparison of related gene copies and DNA sequences is the variability of nucleotide substitution rates [12]. If one takes these properties and the differences in codon bias in different genomes into consideration, divergence rates of gene copies must be individually determined because a single constant rate of nucleotide substitutions for genes does not exist. Still, given such variability of divergence rates for genes within the same genome, deviation from an average rate should be within a statistically significant range to allow comparative analysis between species. For sequences that do not encode functional proteins, like pseudogenes or silent transposable elements, divergence rates are expected to be higher because of the lack of selection. In the case of transposable elements, LTRs are identical when the element is integrated into the genome. Therefore, nucleotide substitution rates for LTRs can be used to estimate a relative insertion time [14]. However, because LTRs are expected to decay faster than genes, it has been proposed to apply a different nucleotide substitution rate for LTRs than genes [15]. Whether or not one can trust these rates in an absolute sense, they can serve as a measure for relating sequences by their degree of homology and pedigree. Therefore, one can classify DNA sequences by homology and positional information.

Orthologous Gene Copies from Sorghum, Rice, and Duplicated Regions of Maize

In order to investigate the degree of conserved gene order and content in the cereals, we recently took advantage of a study of duplicated loci in maize [11] that provided entry points for duplicated chromosomal regions of the maize genome and used them as probes for isolating genomic clones from maize and sorghum [16]. Sequence analysis confirmed that these regions were orthologous because of the preserved order of gene copies. Furthermore, because the entire rice genome was sequenced, orthologous regions from rice were identified computationally. As a consequence, chromosomal regions from the two progenitors of maize were compared with orthologous regions of sorghum and rice. We sequenced five duplicated regions of the maize genome. Because of the lower gene density in maize compared to sorghum, it was necessary to enlarge the window of orthologous regions by identifying overlapping maize BAC clones by chromosome walking. Therefore, 18 BACs with a total length of 2,554 kb were sequenced from maize and 6 BACs with a total length of 681 kb were sequenced from sorghum. These regions aligned with 798 kb of the rice genome. These alignments allowed selection of six additional orthologous gene copies that were linked to the five known loci in maize that served as entry points. Based on a divergence time of ~50 mya for the shared maize and sorghum lineage from the lineage that gave rise to rice, alignment of these 44 orthologous gene copies suggested that the two progenitors of maize and the progenitor of sorghum all diverged about 11.9 mya [12].

Ancestry of Chromosomal Duplications and Chromosome Numbers

Once a physical map of maize inbred B73 was constructed, it was projected onto the genetic map with about 2,000 mapped markers [17]. Because an additional 24,000 sequences were linked to the BAC clones of the physical map, all sequences could be placed in their order on the maize chromosomes [18]. The source of these additional sequences consisted of about 53% unique maize EST sequences, 9% highly conserved genes, and 38% gene-containing BAC-end sequences (BES). This high-density map of gene sequences could then be aligned with the sequenced chromosomes of rice. From this comparison, it became clear that rice contains ancient duplications that have been quadrupled in maize, but there was also a recent duplication in rice that was not further duplicated in maize [18] or in wheat [19]. Therefore, analysis of a small number of grass species has allowed identification of the nature and timing of both segmental duplication and WGD events. Whereas the numerous segmental duplications in the rice lineage occurred over a period of 7–70 million years [20–22] with the largest one of about 3 megabases as recent as 7.7 mya [19], the WGD of maize occurred instantaneously via polyploidization as recently as 4.8 mya [12]. Comparison of rice and maize also suggests that, after WGD, the maize genome temporarily had 20 chromosomes that were subsequently rearranged into 10 larger chromosomes with the loss of 10 centromeres, whereas sorghum maintained its 10 chromosomes [18]. Map-based multi-chromosomal alignments like these have also permitted deconstruction of Brassicaceae and Solanaceae genomes into building blocks of inherited chromosome segments [23, 24].

Paralogous Gene Copies

Alignment of orthologous regions of maize, sorghum, and rice provided another unexpected result. When the sorghum and rice regions were aligned, the number of orthologous gene copies was much higher than in the comparison with the two duplicated regions of maize [16]. In most cases, one of the maize homoeologous regions still contained the collinear copy with sorghum and rice, but the other homoeologous region did not. This finding indicated that after the two progenitors of maize hybridized, one copy of the duplicated genes was deleted from its orthologous position. Therefore, WGD did not ultimately lead to the doubling of gene content in maize relative to sorghum or rice. On the other hand, there is also ample evidence of gene insertions that were not orthologous and gave rise to extra gene copies [25–32]. Because the rice genome has been recently sequenced, it is also possible to search for extra genes in maize and sorghum that were absent in their orthologous positions in rice [16]. In most cases, gene sequences missing from orthologous locations were found to be present in other locations, frequently in more than one copy. This analysis indicated that grass genes are frequently copied and inserted into unlinked positions [33]. Such gene insertions gave rise to interruption of synteny, and could be an important source of raw material for rapid gene evolution, including neofunctionalization [34].

Synteny Blocks and their Uneven Expansion

While the analysis of five orthologous regions of the two progenitors of maize, sorghum, and rice provided a useful first glance at the dynamics of chromosome structure in these grasses, the aligned regions were too small to provide much perspective. The existence of a maize physical map permitted selection of a set of overlapping BAC clones from maize chromosome 1S and maize chromosome 9L that span 7.8 Mb or 17.4 cM and 6.6 Mb or 25.6 cM [35]. These regions were homoeologous to each other and orthologous to a contiguous 4.9 Mb region on rice chromosome 3S. The first dramatic features of this analysis were identifications of the junctions that define synteny blocks; regions of collinearity preserved in both duplicated regions of maize and the rice genome. Based on these synteny blocks, the duplicated maize regions experienced different degrees of LTR-retrotransposon insertion, leading to a differential expansion of these chromosomal segments. Because of the loss of gene duplicates, some maize regions even experienced a contraction, whereas others expanded by a factor of as much as 4.2. Based on the size of the maize genome relative to rice, one would expect an average expansion rate of about 3. Indeed the whole genome alignment between maize and rice as described above found that 1 kb of rice corresponds to about 3.2 kb of maize on average [18]. Therefore, the synteny blocks indicate a local uneven expansion of maize chromosomes relative to rice. Some of the non-collinear junctions consist of sequence blocks that can be as large as a few 100 kb and are sites of inversions. Because these regions were chosen for their density of mapped markers, it is not surprising that they exceed the gene density calculated for one hundred randomly selected clones of the maize genome that were sequenced for comparison [36]. The percentage of coding information in the 100 random regions was estimated to be 8%, which was in close agreement with previous estimates of 7.5% derived from BESs. In these two large sequenced regions, coding space of the region was 12% for chromosome 1S and 16% for chromosome 9L. This high gene density in this region also correlates with a local high recombination frequency. It again confirms that the incongruence between physical and genetic distances is due to the uneven distribution of genic regions in the genome [37].

Another interesting difference between maize and rice is that introns tend to be larger because of the insertion of repeat elements (see also below). As described above, the striking feature is that one of the two duplicated gene copies appears to be lost in the maize genome. Furthermore, missing genes present in maize but absent in rice can be found elsewhere in the rice genome. Although the maize genome had not been sequenced yet, it was possible to ask the reciprocal question because of random maize sequences from gene enrichment libraries, ESTs, and BAC ends. Using these data, one finds that a large fraction of maize genes or gene fragments are present in non-orthologous positions. Therefore, this study further confirms the dynamic process of gene movement in these genomes [35]. Nevertheless, if one compares the AA and the BB genomes of *Oryza* species, which are also derived from a common progenitor like the two progenitors of maize, the degree of divergence is largely confined to non-genic elements [38], suggesting that gene collinearity is more stable if closely related genomes stay as separate species as opposed to undergo hybridization. Therefore, the diploidization of the maize genome might have triggered a greater degree of genome dynamics than would have been observed in the two progenitors maintained as separate species.

Mechanisms of Gene Movement

While it is not surprising to find gene movement after speciation, it was still surprising to discover that apparent gene movement also contributes substantially to haplotype variability [13, 39–43]. There appears to be extensive variability of the same chromosomal regions between different races and inbred lines of maize [43] and even between inbreds derived from the same population [44], suggesting gene movement at high rates in very recent time periods. Haplotype here refers to any unique set of allelic differences that can be inherited as a linked unit. The physical size of such different haplotypes can be quite variable because of the variable expansion of intergenic regions [39, 44, 45]. In contrast to the low level of haplotype variability known from human loci, for instance, haplotype variability in maize includes the presence and absence of apparent genes.

There are basically two mechanisms that create this variability, one is 'cut and paste' and the other is 'copy and paste' [4]. As discussed above, analysis of collinear regions of closely related species, where a gene is collinear in two species or haplotypes but present in an unlinked location in the third species or haplotype, suggests a 'cut and paste' mechanism for gene movement. Alternatively, this apparent gene movement may have arisen from a 'copy and paste' mechanism with a subsequent (perhaps much later) deletion of the orthologous gene copy [46]. Alignment of two haplotypes of maize inbred lines, however, led to a discovery of a 'copy and paste' mechanism that takes place in many instances for gene fragments. When an 'empty' site was compared with one containing gene fragments, junction sequences indicated that the insertion had arisen from the action of a *helitron* [47]. This became clear when the gene fragment was used as a probe to identify the donor site, that was usually a gene missing the *helitron* structural features [41, 42]. Therefore, many apparent non-collinear gene duplications in maize and other plant species are actually gene fragments mobilized by *helitron* transposable elements. Recently, an apparently intact gene has been found copied by a *helitron*, the *ZmCDA3* gene described above [13]. Because the paralogous gene is now located in a new chromosomal environment, its regulation has changed compared to the donor site. Although we do not know the significance of this yet, future genetic analysis of such paralogous gene copies should provide further insights into the neo-functionalization, sub-functionalization, preservation, or loss of genes duplicated by a 'copy and paste' mechanism. However, it is interesting to note that, in the case of a 'copy and paste' mechanism leading to tandem duplication at the *z1C1* locus, the new copies changed their transcriptional regulation and affected the penetrance of a recessive mutation of this transcriptional regulator [33]. Neo-functionalization and sub-functionalization could particularly affect unlinked gene copies that are further diverged [48].

Transposable Elements

The general intensity of genome dynamics has been known for a long time in maize, and was suspected to be largely due to transposable element (TE) action [49]. TEs also move by the two basic mechanisms, 'cut and paste' or 'copy and paste.' Examples of the first are the class II elements or DNA transposable elements; examples of the second are class I elements or retroelements. Because the 'copy and paste' mechanism causes chromosome expansion, one expects that there might be a greater selection against retrotransposition than transposition. For instance, a rice plant derived from tissue culture, where retrotransposition is activated, is often observed to have this class I activity rapidly silenced upon regeneration of plants [50]. On the other hand, the mutable activity of autonomous class II TEs can be maintained over many generations [49].

Because TEs constitute the majority of repeated sequences in the genome, single sequence reads from the ends of cloned genomic DNA provide a good quantitative assessment of TEs in the genome of species that have not been fully sequenced [51–54]. However, a critical step in this analysis is the formation of a repeat library against which to compare end sequences. Although abundant TE families are easily captured from genomic sequences of large-insert clones, the medium and low-copy TE families are harder to extract because they differ between species and the entire genome sequence of one species would not completely predict the repeat content for any other species. Hence, the sequence of the *Arabidopsis* genome yielded few TEs and thus prevented its use as a reference for plant genomes with larger size. An initial survey of the repeat content of the rice nuclear genome resulted in a predicted repeat content of only 10% [51], whereas subsequent analyses suggested a repeat content of ~35% [55].

Another factor in underestimates came from the bias of restriction sites. Maize genomic libraries made from sheared DNA appeared to have a slightly higher percentage of repeat elements (63%) than restricted DNA (58%) with each enzyme generating different preferences [53]. Sequencing 100 random regions of the genome provided an even higher level of repeats, ~66% [36], again probably due to improved repeat collections. As suspected, there is a direct correlation between repeat content and the size of the genome, as shown here between rice and maize. A recent analysis of BAC ends from chromosome 3B of wheat, gave an estimate of 86% for repeat content, consistent with its even larger genome [54].

Class I Elements

Construction of large-insert libraries has been the major source for the analysis of intact class I elements [56–58]. Although it had been known that retroelements contribute to allelic variations from cloning smaller genomic fragments from different strains of the same species [59], it was not until the cloning of large contiguous sequences that it became known that expansion of intergenic regions is enhanced by insertions of retroelements into other retroelements, structures that are also called nested sets of retroelements [57]. Retrotransposition consists of reverse transcription of RNA into DNA and the subsequent integration of the DNA into the genome. A major difference between the transcripts is the presence or absence of LTRs (long terminal repeated sequence). If a poly-A RNA is copied without any trans-acting protein coding regions, the elements are referred to as SINEs (short interspersed repetitive elements). These RNAs are read from RNA polymerase III promoters. If the RNA that is copied contains retrovirus-like coding regions but no LTRs, the elements are referred to as LINEs for long interspersed repetitive elements. Similar elements with LTRs are called LTR-retrotransposons. Concerning the class I elements, mammalian genomes contain predominantly LINEs and SINEs and plant genomes contain predominantly LTR-retrotransposons.

Comparison of orthologous regions between maize, sorghum, and different rice subspecies indicated insertions of both genes and transposable elements that correlate with the overall genome size [29]. When these analyses were extended to include duplicated regions in maize, it was evident that the majority of TEs inserted after the two progenitors of maize diverged [46]. Divergence of LTR sequences also indicated that retrotranspositions occurred in waves after speciation, interestingly in the last few million years [46, 60, 61]. Similar waves of retrotranspositions have also occurred within *Oryza* species that differ two-fold in genome size [62]. A major counteracting force appears to be the decay and removal of retrotransposons and other non-coding sequencing by illegitimate recombination and unequal crossing over between LTRs. This results in chromosome contraction rather than expansion [15, 63]. In addition, recombination or conversion can lead to exchanges between tandemly arranged elements [60]. Analysis of haplotypes from a single region in different inbreds and races of maize indicates that recombination is also the source of new combinations of nested retrotransposon blocks, thereby enhancing the variability of sequence structure and lengths of intergenic regions [4, 43].

In all these accounts, LTR-retrotransposons are the predominant species of repetitive DNA elements. In gene-rich regions of maize, the abundance of one subfamily of LTR retrotransposons, the *copia* elements, often exceeds those of the other subfamily, the *gypsy* elements. This appears to be also true for sequences from BAC ends, 26 versus 21% of total sequence. However, contiguous sequences from 100 random regions of maize indicated that *gypsy* elements are more frequent (30 vs. 21%). Interestingly, a shift in this ratio seems to be confined to regions of chromosomal expansion [35]. Comparison of the synteny blocks for maize chromosome 1S and 9L with rice show a large increase of *copia* elements in expanded regions, which is even true for the 100 random regions when they are compared to their orthologous regions in rice [36]. Among *gypsy* elements, the *Huck* family appears to be the dominant one, while among *copia* elements *Ji* and *Opie* are most abundant, indicating that sequence amplification occurred in waves of particularly active elements [46]. Analysis of EST databases also show that many retroelements are transcribed [53]. Expression of retroelements, even translation, does not necessarily lead to transposition, but might reflect different activities like centromere function [56, 64].

Class II Elements

DNA transposable elements were the first TEs to be discovered because they can yield dramatic somatic phenotypes [49]. In contrast to most class I elements, studied DNA TEs prefer to insert into or near genes and, if these genes have a scorable phenotype, transposition activity can be easily followed [65, 66]. Class II elements can be divided into autonomous and non-autonomous elements. They share the terminal inverted repeats at the 5′ and 3′ ends of the elements [67]. The autonomous element encodes a transposase, the enzyme that is required for the 'cut and paste' mechanism [68]. Insertions are also characterized by target site duplications as for class I elements, but they can vary in length depending on the specific element system involved [69]. Excision of elements is imprecise and can create small in-frame insertions of nucleotides, potentially altering codons and the properties of a protein [70].

Class II elements differ in their mutational activity and developmental timing. Maize *Mutator* elements can exhibit very high mutational activity [71, 72] and elements like them have been referred to as Mu-like elements (MULEs). An interesting property of MULEs appears to be their ability to not only jump into genes but also to acquire gene fragments that become internalized into the MULE. About 3,000 of these 'Pack-MULE' sequences have been observed in the rice genome [73]. However, like the *helitrons* discussed above, they usually contain gene fragments rather than intact genes and it is therefore unclear whether they contribute any gene functions. Interestingly, relative to maize, the rice genome has had far more class II TE activity than class I TE activity. In maize, class I versus II elements represent a relative 95 versus 3% of all repetitive DNA, while in rice these numbers are 65 and 35% [36]. Among DNA TEs in rice, the most abundant elements are Miniature Inverted Repeat Transposable Elements or MITEs, accounting for 14% of all repetitive DNA. In maize, identified MITEs comprise only about 0.4% [35]. For instance, in the large synteny blocks described above, maize has the same average number of MITEs in these regions as rice, whereas the repetitive DNA in the rice chromosome 3S region is about 26% MITEs. MITEs are found close to or within genes, in promoters, untranslated regions and introns [74]. In maize and wheat, the CACTA superfamily represents the greatest mass of class II elements [53, 75]. Although the *Mutator* superfamily appears to be more mutagenic in maize than other transposable elements, it has not amplified to exceedingly high copy numbers. The new class of elements described above, the *helitrons*, appear to amplify by a different mechanism than class I or II elements, as suggested by their homology with rolling circle DNA transposons of bacteria [47]. The *helitrons* appear to represent a small portion of both the rice and maize genomes, but they are harder to detect by computational methods if their internal sequences are replaced by gene fragments [35].

Concluding Remarks

Despite the variability between closely related species [26, 28, 29, 61], the apparent intraspecies violation of gene collinearity in maize was a completely unexpected result [39, 44, 45]. Such divergence between species is less significant because of the lack of interspecies mating. However, when two different cultivars of the same species are crossed, alignment of the two parental chromosomes during meiosis would facilitate crossover between conserved, collinear genes and generate meiotic products that could differ in gene content and collinearity from the two parental chromosomes because each parent might have unique plasticity of paralogous gene copies [4]. Furthermore, non-collinear genes or gene fragments may often be expressed during plant development, so that each parental chromosome in a hybrid could contribute to a unique expression pattern. Indeed, that has been shown to be the case for the *z1C1+BSSS53* and *z1C1+B73* haplotypes [44]. In addition, when the *opaque-2* mutation, which fails to express a transcription factor acting on the *z1C1* locus, is introgressed into BSSS53 and B73, the penetrance of the phenotype differs. While all genes of the *z1C1+B73* haplotype respond to *O2* transcriptional regulation, some of the genes of the *z1C1+BSSS53* haplotype do not [33].

In those cases where apparent non-collinear genes are actually gene fragments, their role is less clear [41, 42]. It is expected that many of these gene fragments could be the raw material for the creation of new genes, perhaps including a large exon shuffling component [76]. In addition, the presence of sense or anti-sense transcripts of gene fragments duplicated in these grass genomes is expected to have high potential for acquiring a regulatory epigenetic role, perhaps as the origin of some microRNAs [77]. In addition, there is now an example of a paralogous sequence of an intact gene, *ZmCDA3*, captured by a *helitron* TE. It is intriguing to envision how paralogous gene copies that are hemizygous in hybrids could contribute to QTLs for many phenotypes, perhaps making major contributions to such vital hybrid characteristics as the heterotic yield boost that is the foundation of the hybrid seed corn industry.

Acknowledgements

Work reported here was supported in part by National Science Foundation Plant Genome Grant #9975618 and #0211851.

References

1 Grant D, Cregan P, Shoemaker RC: Genome organization in dicots: genome duplication in *Arabidopsis* and synteny between soybean and *Arabidopsis*. Proc Natl Acad Sci USA 2000;97: 4168–4173.

2 Blanc G, Hokamp K, Wolfe KH: A recent polyploidy superimposed on older large-scale duplications in the *Arabidopsis* genome. Genome Res 2003;13:137–144.

3 Cannon SB, Mitra A, Baumgarten A, Young ND, May G: The roles of segmental and tandem gene duplication in the evolution of large gene families in *Arabidopsis thaliana*. BMC Plant Biol 2004;4:10.

4 Messing J, Dooner HK: Organization and variability of the maize genome. Curr Opin Plant Biol 2006;9:157–163.

5 Gale MD, Devos KM: Comparative genetics in the grasses. Proc Natl Acad Sci USA 1998; 95:1971–1974.

6 Goodman MM, Stuber CW, Newton K, Weissinger HH: Linkage relationships of 19 enzyme loci in maize. Genetics 1980;96:697–710.

7 Wendel JF, Stuber CW, Edwards MD, Goodman MM: Duplicated chromosomal segments in *Zea mays* L.: Further evidence from Hexokinase isozymes. Theor Appl Genet 1986;72:178–185.

8 Helentjaris T, Weber D, Wright S: Identification of the genomic locations of duplicate nucleotide sequences in maize by analysis of restriction fragment length polymorphism. Genetics 1988;118: 353–363.

9 Ahn S, Tanksley SD: Comparative linkage maps of the rice and maize genomes. Proc Natl Acad Sci USA 1993;90:7980–7984.

10 Moore G, Devos K, Wang Z, Gale MD: Grasses, line up and form a circle. Curr Biol 1995;5: 737–739.

11 Gaut BS, Doebley JF: DNA sequence evidence for the segmental allotetraploid origin of maize. Proc Natl Acad Sci USA 1997;94:6809–6814.

12 Swigonova Z, Lai J, Ma J, Ramakrishna W, Llaca V, Bennetzen JL, Messing J: Close split of sorghum and maize genome progenitors. Genome Res 2004;14:1916–1923.

13 Xu JH, Messing J: Maize haplotype with a helitron-amplified cytidine deaminase gene copy. BMC Genet 2006;7:e52.

14 SanMiguel P, Gaut BS, Tikhonov A, Nakajima Y, Bennetzen JL: The paleontology of intergene retrotransposons of maize. Nat Genet 1998;20:43–45.

15 Ma J, Bennetzen JL: Rapid recent growth and divergence of rice nuclear genomes. Proc Natl Acad Sci USA 2004;101:12404–12410.

16 Lai J, Ma J, Swigonova Z, Ramakrishna W, Linton E, et al: Gene loss and movement in the maize genome. Genome Res 2004;14:1924–1931.

17 Cone KC, McMullen MD, Bi IV, Davis GL, Yim YS, et al: Genetic, physical, and informatics resources for maize. On the road to an integrated map. Plant Physiol 2002;130:1598–1605.

18 Wei F, Coe EH Jr, Nelson W, Bharti AK, Engler FW, et al: Physical and genetic structure of the maize genome reflects its complex evolutionary history. PLoS Genet 2007;3:e3123.

19 Rice-Chromosomes-11-and-12-Sequencing-Consortia: The sequence of rice chromosomes 11 and 12, rich in disease resistance genes and recent gene duplications. BMC Biol 2005;3:20.

20 Vandepoele K, Simillion C, Van de Peer Y: Evidence that rice and other cereals are ancient aneuploids. Plant Cell 2003;15:2192–2202.

21 Paterson AH, Bowers JE, Chapman BA: Ancient polyploidization predating divergence of the cereals, and its consequences for comparative genomics. Proc Natl Acad Sci USA 2004;101: 9903–9908.

22 Yu J, Wang J, Lin W, Li S, Li H, et al: The genomes of *Oryza sativa*: a history of duplications. PLoS Biol 2005;3:e38.

23 Lim KY, Kovarik A, Matyasek R, Chase MW, Clarkson JJ, Grandbastien MA, Leitch AR: Sequence of events leading to near-complete genome turnover in allopolyploid *Nicotiana* within five million years. New Phytologist 2007;175:756–763.

24 Schranz ME, Lysak MA, Mitchell-Olds T: The ABC's of comparative genomics in the Brassicaceae: building blocks of crucifer genomes. Trends Plant Sci 2006;11:535–542.
25 Messing J, Llaca V: Importance of anchor genomes for any plant genome project. Proc Natl Acad Sci USA 1998;95:2017–2020.
26 Tikhonov AP, SanMiguel PJ, Nakajima Y, Gorenstein NM, Bennetzen JL, Avramova Z: Colinearity and its exceptions in orthologous adh regions of maize and sorghum. Proc Natl Acad Sci USA 1999;96:7409–7414.
27 Feuillet C, Penger A, Gellner K, Mast A, Keller B: Molecular evolution of receptor-like kinase genes in hexaploid wheat. Independent evolution of orthologs after polyploidization and mechanisms of local rearrangements at paralogous loci. Plant Physiol 2001;125:1304–1313.
28 Ramakrishna W, Dubcovsky J, Park YJ, Busso C, Emberton J, SanMiguel P, Bennetzen JL: Different types and rates of genome evolution detected by comparative sequence analysis of orthologous segments from four cereal genomes. Genetics 2002;162:1389–1400.
29 Song R, Llaca V, Messing J: Mosaic organization of orthologous sequences in grass genomes. Genome Res 2002;12:1549–1555.
30 Brunner S, Keller B, Feuillet C: A large rearrangement involving genes and low-copy DNA interrupts the microcollinearity between rice and barley at the Rph7 locus. Genetics 2003;164: 673–683.
31 Ilic K, SanMiguel PJ, Bennetzen JL: A complex history of rearrangement in an orthologous region of the maize, sorghum, and rice genomes. Proc Natl Acad Sci USA 2003;100:12265–12270.
32 Langham RJ, Walsh J, Dunn M, Ko C, Goff SA, Freeling M: Genomic duplication, fractionation and the origin of regulatory novelty. Genetics 2004;166:935–945.
33 Song R, Llaca V, Linton E, Messing J: Sequence, regulation, and evolution of the maize 22-kD alpha zein gene family. Genome Res 2001;11:1817–1825.
34 Rodriguez-Trelles F, Tarrio R, Ayala FJ: Convergent neofunctionalization by positive Darwinian selection after ancient recurrent duplications of the xanthine dehydrogenase gene. Proc Natl Acad Sci USA 2003;100:13413–13417.
35 Bruggmann R, Bharti AK, Gundlach H, Lai J, Young S, et al: Uneven chromosome contraction and expansion in the maize genome. Genome Res 2006;16:1241–1251.
36 Haberer G, Young S, Bharti AK, Gundlach H, Raymond C, et al: Structure and architecture of the maize genome. Plant Physiol 2005;139:1612–1624.
37 Dooner HK, Martinez-Ferez IM: Recombination occurs uniformly within the bronze gene, a meiotic recombination hotspot in the maize genome. Plant Cell 1997;9:1633–1646.
38 Kim H, San Miguel P, Nelson W, Collura K, Wissotski M, et al: Comparative physical mapping between *Oryza sativa* (AA genome type) and *O. punctata* (BB genome type). Genetics 2007;176: 379–390.
39 Brunner S, Fengler K, Morgante M, Tingey S, Rafalski A: Evolution of DNA sequence nonhomologies among maize inbreds. Plant Cell 2005;17:343–360.
40 Gupta S, Gallavotti A, Stryker GA, Schmidt RJ, Lal SK: A novel class of Helitron-related transposable elements in maize contain portions of multiple pseudogenes. Plant Mol Biol 2005;57:115–127.
41 Lai J, Li Y, Messing J, Dooner HK: Gene movement by Helitron transposons contributes to the haplotype variability of maize. Proc Natl Acad Sci USA 2005;102:9068–9073.
42 Morgante M, Brunner S, Pea G, Fengler K, Zuccolo A, Rafalski A: Gene duplication and exon shuffling by helitron-like transposons generate intraspecies diversity in maize. Nat Genet 2005;37:997–1002.
43 Wang Q, Dooner HK: Remarkable variation in maize genome structure inferred from haplotype diversity at the bz locus. Proc Natl Acad Sci USA 2006;103:17644–17649.
44 Song R, Messing J: Gene expression of a gene family in maize based on noncollinear haplotypes. Proc Natl Acad Sci USA 2003;100:9055–9060.
45 Fu H, Dooner HK: Intraspecific violation of genetic colinearity and its implications in maize. Proc Natl Acad Sci USA 2002;99:9573–9578.
46 Du C, Swigonova Z, Messing J: Retrotranspositions in orthologous regions of closely related grass species. BMC Evol Biol 2006;6:e62.
47 Kapitonov VV, Jurka J: Rolling-circle transposons in eukaryotes. Proc Natl Acad Sci USA 2001; 98:8714–8719.

48 Dias AP, Braun EL, McMullen MD, Grotewold E: Recently duplicated maize R2R3 Myb genes provide evidence for distinct mechanisms of evolutionary divergence after duplication. Plant Physiol 2003;131:610–620.
49 McClintock B: The significance of responses of the genome to challenge. Science 1984;226:792–801.
50 Hirochika H, Sugimoto K, Otsuki Y, Tsugawa H, Kanda M: Retrotransposons of rice involved in mutations induced by tissue culture. Proc Natl Acad Sci USA 1996;93:7783–7788.
51 Mao L, Wood TC, Yu Y, Budiman MA, Tomkins J, et al: Rice transposable elements: a survey of 73,000 sequence-tagged-connectors. Genome Res 2000;10:982–990.
52 Meyers BC, Tingey SV, Morgante M: Abundance, distribution, and transcriptional activity of repetitive elements in the maize genome. Genome Res 2001;11:1660–1676.
53 Messing J, Bharti AK, Karlowski WM, Gundlach H, Kim HR, et al: Sequence composition and genome organization of maize. Proc Natl Acad Sci USA 2004;101:14349–14354.
54 Paux E, Roger D, Badaeva E, Gay G, Bernard M, Sourdille P, Feuillet C: Characterizing the composition and evolution of homoeologous genomes in hexaploid wheat through BAC-end sequencing on chromosome 3B. Plant J 2006;48:463–474.
55 International-Rice-Genome-Sequencing-Project: The map-based sequence of the rice genome. Nature 2005;436:793–800.
56 Hu W, Das OP, Messing J: Zeon-1, a member of a new maize retrotransposon family. Mol Gen Genet 1995;248:471–480.
57 SanMiguel P, Tikhonov A, Jin YK, Motchoulskaia N, Zakharov D, et al: Nested retrotransposons in the intergenic regions of the maize genome. Science 1996;274:765–768.
58 Chen M, SanMiguel P, de Oliveira AC, Woo SS, Zhang H, Wing RA, Bennetzen JL: Microcolinearity in sh2-homologous regions of the maize, rice, and sorghum genomes. Proc Natl Acad Sci USA 1997;94:3431–3435.
59 Jin YK, Bennetzen JL: Structure and coding properties of Bs1, a maize retrovirus-like transposon. Proc Natl Acad Sci USA 1989;86:6235–6239.
60 Ma J, SanMiguel P, Lai J, Messing J, Bennetzen JL: DNA rearrangement in orthologous orp regions of the maize, rice and sorghum genomes. Genetics 2005;170:1209–1220.
61 Swigonova Z, Bennetzen JL, Messing J: Structure and evolution of the r/b chromosomal regions in rice, maize and sorghum. Genetics 2005;169:891–906.
62 Piegu B, Guyot R, Picault N, Roulin A, Saniyal A, et al: Doubling genome size without polyploidization: dynamics of retrotransposition-driven genomic expansions in *Oryza australiensis*, a wild relative of rice. Genome Res 2006;16:1262–1269.
63 Devos KM, Brown JK, Bennetzen JL: Genome size reduction through illegitimate recombination counteracts genome expansion in *Arabidopsis*. Genome Res 2002;12:1075–1079.
64 Yan H, Ito H, Nobuta K, Ouyang S, Jin W, et al: Genomic and genetic characterization of rice Cen3 reveals extensive transcription and evolutionary implications of a complex centromere. Plant Cell 2006;18:2123–2133.
65 Dooner HK, Belachew A, Burgess D, Harding S, Ralston M, Ralston E: Distribution of unlinked receptor sites for transposed Ac elements from the bz-m2(Ac) allele in maize. Genetics 1994;136:261–279.
66 Cresse AD, Hulbert SH, Brown WE, Lucas JR, Bennetzen JL: Mu1-related transposable elements of maize preferentially insert into low copy number DNA. Genetics 1995;140:315–324.
67 Pohlman RF, Fedoroff NV, Messing J: The nucleotide sequence of the maize controlling element Activator. Cell 1984;37:635–643.
68 Essers L, Adolphs RH, Kunze R: A highly conserved domain of the maize activator transposase is involved in dimerization. Plant Cell 2000;12:211–224.
69 Fedoroff NV: About maize transposable elements and development. Cell 1989;56:181–191.
70 Wessler SR, Baran G, Varagona M, Dellaporta SL: Excision of Ds produces waxy proteins with a range of enzymatic activities. Embo J 1986;5:2427–2432.
71 O'Reilly C, Shepherd NS, Pereira A, Schwarz-Sommer Z, Bertram I, et al: Molecular cloning of the a1 locus of *Zea mays* using the transposable elements En and Mu1. Embo J 1985;4:877–882.
72 Robertson DS, Stinard PS: Genetic analyses of putative two-element systems regulating somatic mutability in *Mutator*-induced aleurone mutants of maize. Dev Genet 1989;10: 482–506.

73 Jiang N, Bao Z, Zhang X, Eddy SR, Wessler SR: Pack-MULE transposable elements mediate gene evolution in plants. Nature 2004;431:569–573.
74 Bureau TE, Wessler SR: Tourist: a large family of small inverted repeat elements frequently associated with maize genes. Plant Cell 1992;4:1283–1294.
75 Wicker T, Guyot R, Yahiaoui N, Keller B: CACTA transposons in Triticeae. A diverse family of high-copy repetitive elements. Plant Physiol 2003;132:52–63.
76 Gilbert W, de Souza SJ, Long M: Origin of genes. Proc Natl Acad Sci USA 1997;94:7698–7703.
77 Ambros V: MicroRNAs: tiny regulators with great potential. Cell 2001;107:823–826.

Joachim Messing
Waksman Institute of Microbiology, Rutgers University
190 Frelinghuysen Road
Piscataway, NJ 08854–8020 (USA)
Tel. +1 732 445 4256, Fax +1 732 445 0072, E-Mail messing@waksman.rutgers.edu

Volff J-N (ed): Plant Genomes.
Genome Dyn. Basel, Karger, 2008, vol 4, pp 57–68

Phylogenetic Insights Into the Pace and Pattern of Plant Genome Size Evolution

C.E. Grover, J.S. Hawkins, J.F. Wendel

Iowa State University, Department of Ecology, Evolution and Organismal Biology, Ames, Iowa, USA

Abstract

It has long been known that organismal complexity is poorly correlated with genome size and that tremendous variation in DNA content exists within many groups of organisms. This diversity has generated considerable interest in: (1) the identity and relative impact of sequences responsible for genome size variation, and (2) the suite of internal mechanisms and external evolutionary forces that collectively are responsible for the observed diversity. Genome size in any given taxon reflects the net effects of multiple mechanisms of DNA expansion and contraction, which by virtue of their complexity and temporal juxtaposition, may be challenging to tease apart into their constituent contributions. Here we review our current understanding of genome size variation in plants and the spectrum of mechanisms thought to be responsible for this variation. We present a synopsis of the insights into the mechanisms and pace of genome size change that are uniquely facilitated by a phylogenetic perspective, particularly among closely related species. We also highlight recent studies in diverse angiosperm groups where comparative genomic approaches have yielded general insights into the myriad mechanisms responsible for much of the observed genome size variation, most prominently the contribution of transposable elements (TEs). Finally, we draw attention to the possibility of divergence in the relative importance of different mechanisms of genome size evolution during cladogenesis.

For more than half a century it has been known that the physical size of a genome does not correlate with organismal complexity [1]. This discrepancy, originally termed the 'C-value paradox', initially was thought to reflect polyploidy or polyteny [2]. It soon became evident, however, that the observed differences in genome size are primarily due to the differential abundance of various kinds of non-coding DNA [3], in a manner independent of variation in ploidy level. During the last several decades, information on genome size has been gathered for thousands of plant species, stimulated by the increasing rapidity with which such information could be gathered, but also by the hope of illuminating the possible biological significance of the observed variation [4]. Paralleling this growth in empirical data has been an increased interest in the directionality of genome size change and the spectrum of mechanisms that might underlie the observed trends [5]. This led to the recognition that genomes are able to contract as well as expand [6], and that the size of any particular genome reflects the net effects of many genetic mechanisms, often acting in opposition with respect to genome size and being subject to and molded by external ecological/evolutionary forces.

Not surprisingly, the most detailed information on genome size variation has emerged from comparisons among agronomically important or model plants for which genomic resources are readily available (e.g. *Arabidopsis*, maize, rice). Studies involving these species have provided important insights into many aspects of genome size variation; however, their great evolutionary distance from one another usually precludes the kind of precise inference one might like to have with respect to the pace of evolutionary change in overall genome size or the relative importance of the various molecular mechanisms that collectively contribute to genome size growth or contraction. A more comprehensive view of genome evolution requires the superimposition of patterns of genomic diversity on an antecedent-descendant framework, as well as an understanding of the temporal dimension to the divergence under study. As in many areas in biology [7], a phylogenetic perspective provides a powerful lens through which the mechanisms that underlie genome evolution may more clearly be brought into focus. A merger of phylogenetics and genomics is likely to lead to a much-improved understanding of long-standing and somewhat vexing questions regarding the mechanisms responsible for shaping genomes as well as the evolutionary transition in their relative importance.

Patterns of Genome Size Variation

Extraordinary variation in genome size has been observed in a wide spectrum of eukaryotes. The variation among land plants as a whole exceeds 2300-fold, with some groups displaying far more variation (pteridophytes; >1300-fold) than others (bryophytes; ~11-fold) [8]. Among angiosperms, the smallest known genome is *Fragaria viridis* (98 Mbp), over 1270 times smaller than the largest known angiosperm genome (the tetraploid *Fritillaria assyriaca*; 124,852 Mbp) and over 890 times smaller than the largest diploid (*Fritillaria davisii*; 87,686 Mbp). That distantly related organisms vary in genome size is no longer unexpected; what is more intriguing is the observation that genome size can vary significantly even among closely related genera or species. Of the 537 genera represented in the Plant C-value database [8] by more than one species, the within-genus variation averages 3-fold; however, the range in genome size variation within genera is from nearly invariant to over 63-fold. An added dimension to the variation pattern is that of intraspecific genome size variation. While some reports of intraspecific genome size variation have later been retracted [9] or are potentially confounded by measurement error [10, 11], intraspecific variation in genome size may be rather common.

The variation noted above permits several inferences regarding evolutionary change in DNA content, even in the absence of explicit phylogenetic analysis or knowledge of underlying mechanism. For example, the phylogenetic placement of plants with small genomes [6, 12, 13], as well as the non-additivity of polyploid genomes with respect to their diploid progenitors [14], suggests that genomes may contract as well as expand [6]. A second implication arises from the observation of the often remarkable variation in genome size among closely related species; namely, that plant genomes may expand or contract rather quickly and that these processes can be accelerated, attenuated, or reversed. The marriage of phylogenetics and genomics is proving particularly fruitful in revealing the evolutionary and mechanistic details regarding this latter set of processes.

Mechanisms that Increase Nuclear DNA Content

Mechanisms that increase the amount of DNA in a nucleus are reasonably well-understood, at least in terms of responsible processes if not the controls on these processes. Polyploidy was one of the first suggested explanations for differences in genome size among plants, although, as it represents the merger of two genomes, it has been subsequently debated whether to consider the whole polyploid genome or the individual genomes that comprise it when considering genome size differences [15–17]. Nevertheless, given the prevalence of polyploidy in plants [18, 19] and that polyploids can subsequently be returned to a diploid state by a process known as 'fractionation' whereby some portion of the duplicate genes are lost [20], polyploidy is clearly an important mechanism by which genomes ultimately expand. Additionally underscoring the importance of polyploidy to genome size evolution is the notion of genome 'down-sizing', a general term describing the lack of additivity of neopolyploid genomes relative to their diploid progenitors [14].

In addition to genome doubling via polyploid formation, the most important mechanism of genomic expansion in plants is transposable element (TE) proliferation. First described in the grasses, the significant contribution of TEs, particularly LTR-retrotransposons, to total nuclear content and microstructure in larger genomes is now widely recognized [21–24]. The ability of TEs to rapidly affect genome size is exemplified by studies in maize, where LTR-retrotransposon proliferation is responsible for doubling the maize genome in as little as three million years [21]. It appears that maize is not exceptional; repetitive sequences, primarily TEs, make up greater than 80% of the total DNA content of angiosperm species with genome sizes greater than 5.0 pg [25]. It is now known that different TE families comprise different proportions of modern genomes, leading to the idea that proliferation of a single or few families of TEs could be responsible for rapid genome growth, and that those TE families need not be the same for all taxa.

Recently, this idea has been demonstrated phylogenetically in *Gossypium*. Using genomic survey sequences representing three *Gossypium* species that differ approximately three-fold in genome size, Hawkins et al. described the spectrum and frequency of sequences that constitute various closely related *Gossypium* genomes [26]. Like maize, TE amplification, specifically proliferation of one particular type of *gypsy*-like LTR-retrotransposon, triggered a 3-fold increase in genome size over the 5–10 my since the origin of the genus. However, the authors demonstrated varying rates of accumulation of different types of TEs in different lineages. For example, LINE-like sequences are relatively inactive in the smallest genome species, but increase in copy number in the species with larger genomes. The converse is seen for accumulation of *copia*-like retrotransposons, which proliferated to the greatest extent in the *Gossypium* species with the smallest genome. It is important to note that these observations were made possible through comparisons of closely related species within a well-established phylogenetic framework.

Smaller-scale mechanisms may stochastically or episodically operate to expand genomes, and these in particular are only directly quantifiable when analyzed within a phylogenetic framework. A recent study in maize [27] provides evidence for a correlation between the large maize genome and increased intron size, relative to rice, which indicates that over large evolutionary distances, variation in intron size may contribute to overall genome size differences. This phenomenon appears to have little, if any, correlation with genome size in plants over a shorter evolutionary distance, as shown by phylogenetic studies in *Gossypium* [28, 29]. The transfer of organellar DNA to the nucleus is another phenomenon that potentially may contribute to nuclear genome expansion, but this has not been evaluated to any appreciable extent within a phylogenetic framework. Studies in rice and *Arabidopsis* indicate such transfers are often represented by plastid 'splinters' [30], which are often small but can represent several hundred kilobases alone [31] or even megabases in toto [32]. Duplication and pseudogenization of genes may, in principle, contribute to genome expansion, but there is little information to suggest that plant lineages vary significantly in the rate of duplication and pseudogenization or that this phenomenon contributes in a substantive way to genome size evolution. Finally, expansion (as well as contraction) of tandem repeat arrays such as rDNA and satellite sequences may also contribute to variation in genome size. Although these mechanisms are not generally thought to be major players in genome size evolution, additional analyses from a phylogenetic perspective may reveal specific examples running counter to this conclusion.

Mechanisms That Decrease Nuclear DNA Content

Mechanisms of DNA loss are not as well understood as those of DNA gain, and thus their relative contributions to global and local patterns of genome size evolution have only recently come under study. The best understood mechanism of DNA loss, intra-strand homologous recombination (recombination between directly repeated sequences), is thought to occur predominantly between the LTRs of single or adjacent retrotransposons [33], a mechanism evidenced by the characteristic hallmark of a solo LTR [34]. Since LTR-retrotransposons often occur in nested arrangements [22, 35], this mechanism has the potential to remove significant amounts of DNA, especially if it involves the LTRs from two spatially separated elements that share high sequence similarity. As shown by Devos et al., the amount of sequence removed via intra-strand homologous recombination can be estimated using the average element length for a given family of LTR-retrotransposons and the involvement of more than one element may be inferred [36]. However, if one is to fully understand the scope and scale of the impact of intra-strand homologous recombination on genome size evolution, it is best inferred when the ancestral condition (single vs. nested elements) has been established. Since LTR-retrotransposons experience considerably rapid turnover (the half-life of LTR-retrotransposons in rice is estimated <6 my [37]), it is necessary to employ closely related taxonomic groups for this type of analysis due to the higher probability of correctly inferring the ancestral state through inclusion of an outgroup.

The suite of mechanisms encompassed by illegitimate recombination (i.e. *RecA* independent recombination involving regions of microhomology) has also been demonstrated to impact genome size [36, 37]. The non-conservative mechanisms of double strand break (DSB) repair, non-homologous end-joining (NHEJ), and slipstrand mispairing are the mechanisms most commonly attributed to illegitimate recombination, although a variety of other mechanisms may be involved. Repair of DSBs through NHEJ appears to operate preferentially in plants [38]. The potential of NHEJ to influence genome size was empirically demonstrated using *Arabidopsis thaliana* with its quintessentially small genome, and *Nicotiana tabacum*, which has a 40-fold larger genome [39, 40]. These studies showed that deletions were larger and more frequent in *A. thaliana* than in *N. tabacum*, and while nearly half of the breakpoints were associated with insertions in tobacco, none were recovered in *Arabidopsis*. Other studies have evaluated the potential of illegitimate recombination to remove DNA from the sequenced *Arabidopsis* and rice genomes, by comparing internally deleted transposable elements with their inferred ancestral sequence [36, 37]. The data suggest that illegitimate recombination may be the 'driving force' behind rapid DNA loss in *Arabidopsis*. Similarly, while second to intra-strand homologous recombination, illegitimate recombination is also largely implicated in DNA loss in rice [37].

A bias in the mechanisms encompassed by illegitimate recombination may be in part responsible for the observation that small genome taxa tend to have more and larger small deletions, which led to the 'mutational equilibrium model' of genome size evolution [41]. This model proposes that genome size shrinks or expands until DNA loss through small deletions is offset by DNA gain through large insertions; organisms with smaller genomes may experience more frequent and larger deletions at equilibrium than those with larger genomes. Currently, limited data exist for the influence of illegitimate recombination on closely related genomes and some have questioned the evolutionary significance of a small indel bias [42]; however, data from *Gossypium* suggests that illegitimate recombination may act more frequently and with longer deletions in species with smaller vs. larger genomes [43]. Whether this is a common phenomenon remains to be seen, but clearly there is much to be learned from further phylogenetic analyses of illegitimate recombination.

Genome Size Increase and Decrease: A Balancing Act

Recent research has shed light on genomic growth through TE amplification and the potential for genome contraction through intra-strand homologous recombination and illegitimate recombination. Through characterization of random genomic sequences, Piegu et al. [44] demonstrated lineage-specific amplification of retrotransposons within the last two million years in *Oryza australiensis*, providing further evidence for major increases in genome size via recent TE amplification. To investigate the potential for DNA loss through intra-strand homologous recombination, the authors also estimated the number of solo-LTRs belonging to those same TE families via slot blots. Additionally, the authors sequenced several hundred kilobases of contiguous genomic sequence to gauge the effect of illegitimate recombination on genome size differences. This study demonstrated that while unequal homologous recombination and illegitimate recombination are operating in *Oryza australiensis,* unequal homologous recombination is far more efficient in this genome, often occurring concurrently with TE amplification by removing a fraction of the transposed elements immediately upon or shortly after the transpositional burst. In contrast, illegitimate recombination appears to have a negligible effect on DNA loss in this species.

In a separate study, Vitte and Bennetzen [45] analyzed several megabases of genomic sequence to investigate TE composition for five diverse angiosperm genomes and compared that to the published data for rice and *Arabidopsis*. The authors were able to determine the relative importance of intra-strand homologous recombination and illegitimate recombination in each genome by comparing the number of truncated TEs (products of illegitimate recombination) with the number of solo-LTRs for that genome. This ratio varied by more than two-fold, leading to the conclusion that each acts with different efficiencies uncorrelated with phylogenetic relatedness or genome size [45]. An additional observation was that the relative abundance of intact elements does not correlate with genome size. *Hordeum* (1C = 5400 Mb) and *Medicago* (1C = 470 Mb), for instance, contain the lowest percentage of intact sequences, while a second monocot with a large genome (maize, 1C = 2700 Mb) and a second dicot with a small genome (*Lotus*, 1C = 470 Mb) contained the highest percentage of intact elements. They conclude that while lineage-specific differences also exist for the aggressiveness of the primary mechanisms of DNA removal, they may be less significant in the determination of genome size differences on a broad scale than TE amplification. Comparative studies of orthologous genomic regions within a close phylogenetic framework will be necessary to determine, from a quantitative standpoint, the relative strengths of mechanisms that operate to add versus remove DNA from a genome on a finer scale.

The aforementioned studies highlight in part the vibrancy of the field of genome size evolution. With the data generated from genome sequencing projects and low-cost, large scale sequencing for other genomes, researchers are beginning to understand the suite of mechanisms responsible for genome size evolution and the impact each has on genome size, leading to several general insights. First, transposable element amplification is a primary contributor to genome size growth, having the potential to dramatically increase genome size over a short period of time. These episodic TE bursts represent the amplification of different families in different lineages at different time points, which is reflected by the age and abundancy of TE families in related and distant genomes. Second, growth by TE proliferation is counteracted (to varying degrees) by two prominent mechanisms of DNA removal, intra-strand homologous recombination and illegitimate recombination, which operate with greater or lesser strength in different genomes, and may be uncorrelated with genome size or phylogenetic placement. Third, mechanisms that potentially could operate throughout the genome (e.g. deletional bias and TE proliferation), appear to affect different regions differently [29] and affect similar regions dissimilarly [27].

Genome Size and the Organism

Discussions of genome size variation often focus on mechanisms of expansion and contraction, as we have here. It needs to be remembered, however, that these myriad mechanisms operate in an organismal and ecological context involving multiple levels of evolutionary constraint and pressure. A full accounting of these is beyond the scope of this review, and in fact may not even be possible, but minimally one would wish to know the effects of genome size on nuclear volume, the cell cycle, energetic balance, optimal rates of growth and development, and ultimately reproductive fitness. At present very little is truly understood about most of these connections between adaptation and genomic organization, although a great deal of attention has been given to possible correlations between genome size and various life-history features. These include traits such as cell volume [3, 46, 47], endopolyploidy [48], leaf size [49–52], annual/weedy lifestyle [53–55], cell cycle duration [56], drought tolerance [49, 51, 57], frost tolerance [58], and altitude [59].

Many of the foregoing correlations have been in conflict (positive and negative correlations for the same trait), leading some to suggest that plants with large genomes are excluded from extreme environments [60]. A common criticism of such correlations is that they are not readily verifiable through experimentation [61]. This probably is true in most cases, but as more is learned about the mechanistic underpinnings of the observed patterns, hypotheses concerning how mechanisms relate to changes in biology/ecology may become apparent. An illustrative case in point is offered by wild barley (*Hordeum spontaneum*) from 'Evolution Canyon' in Israel [62]. In this remarkable study, Kalendar et al. investigate transposable element copy number in individuals relative to their ecogeographical distribution and local microclimate. The data suggest the responsiveness of the BARE-1 *copia* retroelement to increased stress (aridity) in the different microclimates; that is, increased stress was associated with more BARE-1 copies and proportionally fewer solo-LTRs. The authors suggest this reflects increased amplification and retention of BARE-1 elements, ultimately creating larger genomes in the more stressful microclimates.

One potential criticism of the wild barley work and correlative studies is that the polarity of genomic change (e.g., insertion vs. deletion) is often unknown. Explicitly phylogenetic approaches may be particularly promising in this respect, as genomic characters and transformations associated with ecological and life-history variables may be mapped onto organismal phylogenies and thereby possibly reveal interactions and associations across levels of biological organization. Patterns may emerge that implicate genome size itself as a trait with specific evolutionary effects. Conversely, genome size itself may not be ecologically or evolutionarily relevant; instead, it may reflect the aggregate effects of internal molecular mechanisms, a portion of which may be environmentally influenced. For example, it is well known that hybridization and environmental stresses may lead to transposable element proliferation [63–65], some portion of which may mediate adaptation via insertional mutagenesis or effects on regulatory regions. One may readily envision situations where TE proliferation leads to both a vastly expanded genome and only one or a few adaptively relevant insertions. In these cases genome size per se may be selectively irrelevant, a mere passive hitch-hiker on favorable mutations, yet it might be correlated with some other ecological feature or trait of interest. At present, the interactions among genome size change, molecular genetic mechanisms, and biological/ecological traits are not understood, although the rapidly expanding knowledge base and increasing experimental capabilities in some systems suggest that we are entering an era in which these elusive questions may be experimentally addressed.

Future Directions

Our understanding of the mechanisms responsible for genome size evolution has vastly improved in the past decade. The pattern of recurring, episodic lineage-specific amplification of transposable elements (both with respect to plant lineages and TE family) begs the question of why particular TE families are able to successfully colonize some genomes versus others. What are the specific familial attributes or selective advantages that permit, prohibit, or promote proliferation? Detailed analysis of the dominant class/family of TEs in a genome and the shifts that have occurred from one class/family to another will shed light on the permissive, or perhaps complementary, properties of successful TEs and the genomes they inhabit. Similarly, deletional mechanisms, whose initial obscurity prompted the notion of a 'one-way ticket to genomic obesity' [5], are now known to have the ability to remove substantial amounts of DNA from a genome. The question remains as to whether these mechanisms are able to shrink a genome over time or whether they mainly function to slow genome growth. Much remains to be learned concerning the nature of and internal and external controls of deletional mechanisms and their links to double-stranded break repair, which, when better understood, will elucidate further their contribution to genome size evolution. As discussed here, bringing a phylogenetic perspective to bear on these questions promises many new insights in the coming years into the mechanisms and evolutionary forces shaping plant genomes.

References

1 Mirsky AE, Ris H: The DNA content of animal cells and its evolutionary significance. J Gen Physiol 1951;34:451–462.

2 Thomas CA: The genetic organisation of chromosomes. Annu Rev Genet 1971;5:237–256.

3 Cavalier-Smith T: Cell volume and the evolution of eukaryote genome size; in Cavalier-Smith T (ed): The Evolution of Genome Size. John Wiley, New York, 1985, pp 104–184.

4 Bennett MD, Leitch IJ: Plant genome size research: A field in focus. Ann Bot 2005;95:1–6.

5 Bennetzen JL, Kellogg EA: Do plants have a one-way ticket to genomic obesity? Plant Cell 1997;9:1509–1514.

6 Wendel JF, Cronn RC, Johnston JS, Price HJ: Feast and famine in plant genomes. Genetica 2002;115:37–47.

7 Avise JC: Evolutionary Pathways in Nature: A Phylogenetic Approach. Cambridge University Press, New York, 2006.

8 Bennett MD, Leitch IJ: Plant DNA C-values database (release 4.0, october 2005). http://wwwrbgkeworguk/cval/homepagehtml 2005.

9 Price HJ, Johnston JS: Influence of light on DNA content of *Helianthus annuus*. Proc Natl Acad Sci USA 1996;93:11264–11267.

10 Greilhuber J: Intraspecific variation in genome size in angiosperms: Identifying its existence. Ann Bot 2005;95:91–98.

11 Noirot M, Barre P, Duperray C, Hamon S, De Kochko A: Investigation on the causes of stoichiometric error in genome size estimation using heat experiments: Consequences on data interpretation. Ann Bot 2005;95:111–118.

12 Bennett MD, Leitch IJ: Nuclear DNA amount in angiosperms. Philos Trans R Soc Lond B Biol Sci 1997;334:309–345.

13 Leitch IJ, Soltis DE, Soltis PS, Bennett MD: Evolution of DNA amounts across land plants (Embryophyta). Ann Bot 2005;95:207–217.

14 Leitch IJ, Bennett MD: Genome downsizing in polyploid plants. Biol J Linnean Soc 2004;82: 651–663.

15 Bennett MD, Leitch IJ: Genome size evolution in plants; in Gregory TR (ed): The Evolution of the Genome. Elsevier academic press, Burlington, MA, 2005, pp 89–162.

16 Gregory TR: Genome size evolution in animals; in Gregory TR (ed): The Evolution of the Genome. Elsevier academic press, Burlington, MA, 2005, pp 3–87.

17 Greilhuber J, Dolezel J, Lysak MA, Bennett MD: The origin, evolution and proposed stabilization of the terms 'genome size' and 'C-value' to describe nuclear DNA contents. Ann Bot 2005;95: 255–260.

18 Tate JA, Soltis DE, Soltis PS: Polyploidy in plants; in Gregory TR (ed): The Evolution of the Genome. Elsevier academic press, Burlington, MA, 2004, pp 371–425.

19 Adams KL, Wendel JF: Polyploidy and genome evolution. Curr Opin Plant Biol 2005;8:135–141.

20 Ilic K, SanMiguel PJ, Bennetzen JL: A complex history of rearrangement in an orthologous region of the maize, sorghum, and rice genomes. Proc Natl Acad Sci USA 2003;100:12265–12270.

21 SanMiguel P, Bennetzen JL: Evidence that recent increase in maize genome size was caused by the massive amplification of intergene retrotransposons. Ann Bot 1998;82:37–44.

22 SanMiguel P, Tikhonov A, Jin YK, Motchoulskaia N, Zakharov D, et al: Nested retrotransposons in the intergenic regions of the maize genome. Science 1996;274:765–768.

23 Chen M, SanMiguel P, de Oliveira AC, Woo S-S, Zhang H, et al: Microcolinearity in *sh2*-homologous regions of the maize, rice, and sorghum genomes. Proc Natl Acad Sci USA 1997;94: 3431–3435.

24 Chen M, SanMiguel P, Bennetzen JL: Sequence organization and conservation in *sh2/a1*-homologous regions of sorghum and rice. Genetics 1998;148:435–443.
25 Flavell RB, Bennett MD, Smith JB, Smith DB: Genome size and the proportion of repeated nucleotide sequence DNA in plants. Biochem Genet 1974;12:257–269.
26 Hawkins JS, Kim H, Nason JD, Wing RA, Wendel JF: Differential lineage-specific amplification of transposable elements is responsible for genome size variation in *Gossypium*. Genome Res 2006;16:1252–1261.
27 Bruggmann R, Bharti AK, Gundlach H, Lai J, Young S, et al: Uneven chromosome contraction and expansion in the maize genome. Genome Res 2006;16:1241–1251.
28 Wendel JF, Cronn RC, Alvarez I, Liu B, Small RL, Senchina DS: Intron size and genome size in plants. Mol Biol Evol 2002;19:2346–2352.
29 Grover CE, Kim H, Wing RA, Paterson AH, Wendel JF: Incongruent patterns of local and global genome size evolution in cotton. Genome Res 2004;14:1474–1482.
30 Shahmuradov IA, Akbarova YY, Solovyev VV, Aliyev JA: Abundance of plastid DNA insertions in nuclear genomes of rice and *Arabidopsis*. Plant Mol Biol 2003;52:923–934.
31 Stupar RM, Lilly JW, Town CD, Cheng Z, Kaul S, et al: Complex mtDNA constitutes an approximate 620-kb insertion on *Arabidopsis thaliana* chromosome 2: Implication of potential sequencing errors caused by large-unit repeats. Proc Natl Acad Sci USA 2001;98:5099–5103.
32 Richly E, Leister D: NUPTs in sequenced eukaryotes and their genomic organization in relation to NUMTs. Mol Biol Evol 2004;21:1972–1980.
33 Shirasu K, Schulman AH, Lahaye T, Schulze-Lefert P: A contiguous 66-kb barley DNA sequence provides evidence for reversible genome expansion. Genome Res 2000;10:908–915.
34 Lim JK, Simmons MJ: Gross chromosome rearrangements mediated by transposable elements in *Drosophila melanogaster*. BioEssays 1994;16:269–275.
35 SanMiguel P, Gaut BS, Tikhonov A, Nakajima Y, Bennetzen JL: The paleontology of intergene retrotransposons in maize. Nat Genet 1998;20:43–45.
36 Devos KM, Brown J, Bennetzen JL: Genome size reduction through illegitimate recombination counteracts genome expansion in *Arabidopsis*. Genome Res 2002;12:1075–1079.
37 Ma J, Devos KM, Bennetzen JL: Analyses of LTR-retrotransposon structures reveal recent and rapid genomic DNA loss in rice. Genome Res 2004;14:860–869.
38 Gorbunova V, Levy AA: How plants make ends meet: DNA double-strand break repair. Trends Plant Sci 1999;4:263–269.
39 Orel N, Puchta H: Differences in the processing of DNA ends in *Arabidopsis thaliana* and tobacco: Possible implications for genome evolution. Plant Mol Biol 2003;51:523–531.
40 Kirik A, Salomon S, Puchta H: Species-specific double-strand break repair and genome evolution in plants. EMBO J 2000;19:5562–5566.
41 Petrov DA: Mutational equilibrium model of genome size evolution. Theor Popul Biol 2002;61: 531–544.
42 Gregory TR: Insertion–deletion biases and the evolution of genome size. Gene 2004;324:15–34.
43 Grover CE, Kim HR, Wing RA, Paterson AH, Wendel JF: Microcolinearity and genome evolution in the *AdhA* region of diploid and polyploid cotton (*Gossypium*). Plant J 2007;50:995–1006.
44 Piegu B, Guyot R, Picault N, Roulin A, Saniyal A, et al: Doubling genome size without polyploidization: Dynamics of retrotransposition-driven genomic expansions in *Oryza australiensis*, a wild relative of rice. Genome Res 2006;16:1262–1269.
45 Vitte C, Bennetzen JL: Analysis of retrotransposon structural diversity uncovers properties and propensities in angiosperm genome evolution. Proc Natl Acad Sci USA 2006;103: 17638–17643.
46 Cavalier-Smith T: Economy, speed and size matter: Evolutionary forces driving nuclear genome miniaturization and expansion. Ann Bot 2005;95:147–175.
47 Cavalier-Smith T: Skeletal DNA and the evolution of genome size. Annu Rev Biophys Bioeng 1982;11:273–302.
48 Nagl W: DNA endoreduplication and polyteny understood as evolutionary strategies. Nature 1976;261:614–615.
49 Castro-Jimenez Y, Newton R, Price HJ, Halliwell R: Drought stress responses of *Microseris* species differing in nuclear DNA content. Am J Bot 1989;76:789–795.

50 Chung J, Lee JH, Armumuganathan K, Graef GL, Specht JE: Relationships between nuclear DNA content and seed and leaf size in soybean. Theor Appl Genet 1998;96:1064–1068.
51 Wakamiya I, Newton R, Johnston JS, Price HJ: Genome size and environmental factors in *Pinus*. Am J Bot 1993;80:1235–1241.
52 Ceccarelli M, Minelli S, Falcinelli M, Cionini PG: Genome size and plant development in hexaploid *Festuca arundinaceae.* Heredity 1993;71:555–560.
53 Chase MW, Hanson L, Albert VA, Whitten WM, Williams NH: Life history evolution and genome size in subtribe Oncidiinae (Orchidaceae). Ann Bot 2005;95:191–199.
54 Bennett MD: Variation in genomic form in plants and its ecological implications. New Phytol 1987;106:177–200.
55 Albach DC, Greilhuber J: Genome size variation and evolution in *Veronica.* Ann Bot 2004;94: 897–911.
56 Bennett MD: Nuclear DNA content and minimum generation time in herbaceous plants. Proc R Soc Lond B Biol Sci 1972;181:109–135.
57 Wakamiya I, Price HJ, Messina M, Newton R: Pine genome size diversity and water relations. Physiol Plant 1996;96:13–20.
58 MacGillivray C, Grime J: Genome size predicts frost resistance in British herbaceous plants: Implications for rates of vegetation response to global warming. Funct Ecol 1995;9:320–325.
59 Knight CA, Ackerly DD: Variation in nuclear DNA content across environmental gradients: A quantile regression analysis. Ecol Lett 2002;5:66–76.
60 Knight CA, Molinari NA, Petrov DA: The large genome constraint hypothesis: Evolution, ecology and phenotype. Ann Bot 2005;95:177–190.
61 Bennetzen JL, Ma J, Devos KM: Mechanisms of recent genome size variation in flowering plants. Ann Bot 2005;95:127–132.
62 Kalendar R, Tanskanen J, Immonen S, Nevo E, Schulman A: Genome evolution of wild barley (*Hordeum spontaneum*) by BARE-1 retrotransposon dynamics in response to sharp microclimatic divergence. Proc Natl Acad Sci USA 2000;97:6603–6607.
63 Wessler SR: Plant retrotransposons: Turned on by stress. Curr Biol 1996;6:959–961.
64 Capy P, Giuliano G, Biemont C, Bazin C: Stress and transposable elements: Co-evolution or useful parasites? Heredity 2000;85:101–106.
65 McClintock B: The significance of responses of the genome to challenge. Science 1984;226: 792–801.

Jonathan F. Wendel
253 Bessey Hall, Department of Ecology, Evolution and Organismal Biology
Iowa State University
Ames, IA 50011 (USA)
Tel. +1 515 294 7172, Fax +1 515 294 1337, E-Mail jfw@iastate.edu

Volff J-N (ed): Plant Genomes.
Genome Dyn. Basel, Karger, 2008, vol 4, pp 69–82

Plant Transposable Elements

J.M. Deragon[a,c], *J.M. Casacuberta*[b], *O. Panaud*[c]

[a]CNRS UMR6547 BIOMOVE, Université Blaise Pascal, Aubière, France; [b]Departament de Genetica Molecular, Laboratori de Genetica Molecular Vegetal, CSIC-IRTA, Barcelona, Spain; [c]Laboratoire Genome et Développement des plantes, Université de Perpignan, Perpignan, France

Abstract

Genomic programs are yielding tremendous amounts of data about plant genomes and their expression. In order to exploit and understand this data it will be necessary to determine the mechanisms leading to natural variation of patterns of gene expression. The ability to understand how gene expression varies among populations (and not only within the population used in the genomics program) and following the exposure of plants to various stress conditions will be fundamental to progress in the post-genomics phase. Transposable elements (TEs) make up nearly half of the total amount of DNA in many plant genomes, so definition of their influence on genome structure and gene expression is of clear significance to the understanding of global genome regulation and phenotype variations. We describe here the different types of plant TEs and recent examples on how they contribute to structure, evolution and genetic control architecture of plant genomes.

Transposable elements (TEs) are defined as DNA sequences able to move from one genomic position to another in a replicative or non-replicative process. Many of them (the autonomous elements) encode protein(s) required for their mobility and use the host cellular machinery for their transcription and translation. In contrast, the non-autonomous elements must use *in trans* the enzymatic machinery of an autonomous element for transposition. TEs are usually classified in two major groups: class I elements that use an RNA intermediate and a reverse transcriptase and class II elements that use a DNA intermediate and a transposase. The class I elements are the most abundant ones in plants. They include the autonomous LINEs (Long Interspersed Elements), the non-autonomous SINEs (Short Interspersed Elements) and the LTR (Long Terminal Repeat) retrotransposons. Non-autonomous LTR-retrotransposons can be further divided in the TRIM (Terminal-repeat Retrotransposons in Miniature) and LARD (LArge Retrotransposon Derivatives) families. The class II elements are composed of autonomous and non-autonomous Transposon, Helitron and Foldback elements as well as the non-autonomous MITEs (Miniature Inverted-repeated Transposable Elements).

At the end of the 1970s and the beginning of the 1980s, TEs were generally considered as 'selfish genes', suggesting that these elements spread by forming additional copies of themselves within the host genome, and did not contribute to the phenotype. However, this vision has considerably changed over the last few years with a growing body of evidence that although TEs can be highly deleterious for the hosts, they can also be beneficial since new genetic structure and regulation may appear favoring a better adaptation to environmental conditions. Consequently, it is generally admitted that TEs can stimulate genome reorganization (e.g., by genomic duplication, inversion or deletion resulting from unequal crossover), promote the emergence of new transcriptional regulation pathways (by associating genes to new promoters, or new regulatory elements, by transcriptional interference, or by modifying the methylation status of a promoter), and generate variability at the post-transcriptional level (by introducing new splicing, polyadenylation, and editing signals in pre-mRNAs). We will briefly review below data on the different plant TEs and focus on newly described examples of the influence of TEs on plant genomes.

Class I: LTR Retrotransposons

LTR-retrotransposons (LTR-RTs) are class I TEs that can be distinguished from LINEs and SINEs by the presence of long terminal sequence, repeated in the same orientation at both ends of the element (i.e. the LTRs). LTR-RTs are found in the genome of all eukaryotes, although they are particularly abundant in the genomes of higher plants (both gymnosperms [1] and angiosperms [2]) compared to animals or lower eukaryotes. The size of LTRs can vary greatly from one family to the other, but usually ranges from several hundreds to several thousands of base pairs. Full length autonomous LTR-RTs contain a gene encoding the GAG (involved in the production of the virus-like particle) and POL (all enzymatic activities, including the reverse transcriptase) functions needed for retrotransposition. Some LTR-RT families in plants harbor, in addition to GAG-POL, a domain encoding a putative envelope-like protein, referred to as the ENV domain. Non-autonomous LTR-RTs are of two types: TRIMs [3] and LARDs [4]. TRIMs are identical in structure to LTR-RTs, although their size does not exceed few hundred base pairs and their internal region is not homologous to any known GAG-POL sequence. LARDs are also identical in structure to LTR-RTs and usually possess LTRs closely related to the LTRs of an active element. However, as for TRIMs, LARDs internal region lacks homology with any known GAG-POL sequence [4]. Taking into account the autonomous and non-autonomous elements, the overall size of LTR-RTs in plants usually ranges from 0.5 to 15 thousand base pairs (kbp) [5]. LTR-RTs are generally scattered throughout the genome of plants, although several studies have shown they more densely populate pericentromeric regions [5] and can be present in centromeres, interspersed between clusters of tandemly arrayed satellite DNA [6].

LTR-Retrotransposons Are Major Actors of Plant Genome Structure and Evolution

Consequently to their relatively large size (compared to other TEs) and propensity to amplify to large copy number, LTR-RTs impact strongly on plant genome structure and evolution. Since the discovery that LTR-RTs compose at least 50% of maize genome [7], many studies have evidenced that they are indeed the main components of large plant genomes [2]. It is in addition now clearly established that, in angiosperms, genome size correlates with the overall genomic content of LTR-RTs, suggesting that they are an important cause of genome size variations [8].

Over the past few years, several studies have focussed on the actual process through which LTR-RTs contribute to plant genome evolution, more precisely the timing and extent of their transpositional activity on the one hand and the fate of the elements posterior to their insertion on the other. Many of these studies have been conducted on the model species *Oryza sativa* (the Asian cultivated rice), mainly because the genome of this species has been entirely sequenced, thus allowing for the first time to perform genome-wide extensive surveys of LTR-RTs in a plant species [9]. The results of these analyses can be synthesized as follows: (i) The highly repeated LTR-RTs originate from transpositional bursts, i.e. from a strong retrotranspositional activity spanning short time periods, rather than from a continuous process. Some recent reports have shown that these bursts of retrotransposition can yield several thousand and even tens of thousands of new copies so that the activity of a single or few LTR-RT families can have a significant impact on genome size variation within a plant species [10, 11]. (ii) These bursts are usually recent (not older than 5 my) [12, 13], and (iii) LTR-RTs are very efficiently eliminated from the genome by both recombinations and deletions (for example, the half-life of LTR-RTs in rice genome has been estimated to be 2 million years [14]). From these three points, a general model of plant genome evolution related to the activity of LTR-RTs has been proposed [5, 14]. It postulates that the intergenic regions, mainly composed of LTR-RTs, evolve through the combined effect of two counteracting forces (i.e. the retrotransposition and the recombination/deletion), which leads to a rapid turn-over of this genomic compartment. One consequence of this is a rapid loss of synteny in intergenic regions in plant genomes. This is particularly well illustrated in the Poaceae family where comparative genomic studies among cereal species (mainly rice, maize, sorghum and wheat) show a complete lack of correspondence of intergenic regions, while, to some extent, gene content and gene order are often observed in orthologous loci [15]. Some recent studies have shown that other classes of TEs can also contribute to a significant extent to the differentiation of plant genomes through a process that is different from that of LTR-RTs. For example, Helitrons have been shown to be at the origin of gene movements in maize (see below).

Impact of LTR-Retrotransposons on Gene Expression

The next steps towards our full understanding of the impact of LTR-RTs on genome biology will certainly consist in characterizing the functional relationships between transposable elements and the genes of their host genome. Recently, the transcription of LTR-retrotransposons was shown to be activated in synthetic hexaploid wheat and that this activation affects the expression of wheat genes [16]. The expression of plant LTR-RTs is often induced in stress situations, and it has been shown that their LTRs can harbor stress-inducible promoters [17, 18]. The induction of the promoters located within the LTR can influence adjacent genes, either by interfering with their promoter or via the production of hybrid transcripts extending into adjacent genes from LTR sequences. Moreover, solo-LTRs (originating from the ectopic recombination between the two LTRs of a given element) can also putatively affect the expression of nearby genes in case they harbor an intact promoter. This has been demonstrated in animal genomes, but still remains to be evidenced for plants.

Class I: LINEs

LINEs are an abundant class of mobile genetic sequences that insert into new genomic locations via reverse transcription of an RNA intermediate by a process termed retroposition. LINE retroposition requires element-encoded endonuclease, reverse transcriptase and nucleic acid binding activities. LINEs apparently arose early in evolution as they are present in genomes of almost all eukaryotes including plants (for a review on LINE see [19]). Full length LINE elements range in size from 4 to 9 kbp but most LINEs in genomes are 5′-truncated copies and are therefore much smaller. The first plant LINE has been identified as an insertion in the 3′-untranslated region of the maize *A1* gene and was named *Cin-4* [20]. Based on conserved regions from the reverse transcriptase gene, LINEs have been subsequently found in many species across the plant kingdom including monocots and dicots [21]. In most cases, plant LINEs are present at low (<100) to moderate (up to 1000) copy numbers in genomes with the exception of the *del2* family from *Lilium* species [22]. *del2* is unusual by its extreme abundance (>2.5×10^5 copies in *Lilium speciosum* representing 4% of the genome) and by the fact that most copies are full-size in contrast to the truncated nature of LINEs. Only two plant LINEs with retroposition capability have been characterized to date, the *Karma* element from rice [23] and the *LIb* element from sweet potato [24].

In mammals, LINEs have been shown to participate actively in the regulation of gene expression [25] while in *Drosophila* they are involved in the telomeric function [26]. At this time, nothing is known about the impact of plant LINEs on gene expression.

Class I: SINEs

SINEs are small (80 to 500 base pairs long) non-autonomous retroelements that use the enzymatic machinery of autonomous LINEs for retroposition [27, 28]. In primates, the dominant SINEs are ancestrally related to 7SL RNA while, for other eukaryotes, SINEs derived from tRNAs are dominant [29]. SINEs genomic copy number usually ranges from a few hundred to more than a million copies for the human *Alu* family [29]. SINEs have been identified in many plant families including Gramineae, Commelinaceae, Rosaceae, Solanaceae, Fabaceae and Brassicaceae [30–34]. All plant SINEs share key characteristics including a tRNA origin, an internal polymerase III promoter, a short stretch of T or A at their 3′-end and the presence of flanking direct repeats. Plant SINEs are mainly dispersed randomly in genomes although they are rarely present in heterochromatic, pericentromeric regions, and have a preference for gene-rich regions. The presence or absence of a SINE at a given locus is an information that can be used to design highly informative molecular markers (reviewed in [35], see [36, 37] for examples of the use of SINE markers in rice and Brassicaceae).

The first SINE described in plants, *p-SINE1*, was found in the *Oryza sativa* (rice) *Waxy* gene [32]. The copy number of *p-SINE1* in rice is estimated to be 6500 per haploid genome [38]. Several other rice sequences, annotated as SINEs, have been recently deposited in REPBASE [39] and it is therefore likely that many other rice SINE families remain to be characterized. The *TS* SINE family has been found in the genomes of several Solanaceae and in one Convolvulaceae species [34, 40]. The copy number of *TS* element in tobacco is estimated to be 5×10^4 per haploid genome. The *AU* SINE family shows the broadest distribution in plants, with related elements being found in Gramineae, in Fabaceae, and in Solanaceae species [33]. The highest copy number is found in *Aegilops umbellulata* and

Triticum aestivum (around 1×10^4 copies) where the element underwent a recent explosive increase in its copy number [33]. The first Brassicaceae SINE family (called *SB1* (formerly *S1*)) [31] was identified in *Brassica napus* [41]. Fifteen different SINE families, with maximum copy numbers around 1×10^3 copies, have now been identified in Brassicaceae including six families in *Arabidopsis thaliana* (for a review see [42]). The current distribution of known SINE families suggests that SINEs are ubiquitous in plants and that the number of characterized plant SINE families will increase with the number of sequenced plant genomes.

SINE RNAs as Regulators of Essential Cellular Function?

Several recent experimental results suggest that SINE polIII specific transcripts could, in stress situations, play a role as non-coding riboregulators (see [43] and references therein). In mammals, SINE RNAs have been shown to regulate translation by interacting with the *PKR* kinase (an enzyme that phosphorylates *EIF2α* and downregulates translation), or by a *PKR*-independent mechanism or to regulate transcription by binding to the *RPB1* subunit of the RNA polymerase II complex. In this context, mammalian SINEs can be considered 'stress responsive-genes' capable of modulating the level of mRNAs and proteins in different stress situations. Nothing is known about the impact of SINE RNAs in plants but the expression of a SINE from *Brassica* in *Arabidopsis* is associated with severe developmental defects and partial sterility (J.M. Deragon, unpublished data), a result compatible with the hypothesis of a riboregulator function for plant SINE RNAs.

Class II: Transposons and MITEs

Class II transposons are characterized by the presence of terminal inverted repeats (TIRs) and the capacity to encode a transposase that catalyses their mobilization, although defective elements without a transposase coding capacity also exist. These elements can only transpose when mobilized *in trans* by their autonomous counterparts. Most defective Class II transposons are mutation derivatives of autonomous elements and they share with the later extensive sequence identity. However, defective elements with very limited sequence similarity to autonomous elements also exist. The TIR and subterminal sequences and, in particular the transposase binding sites, are usually the only essential sequences for mobilization. After binding the TIRs, different units of transposase contact each other joining the ends of the element to generate a particular structure known as transpososome, which allows the transposase to cleave the DNA and excise the transposon. The transposase is also responsible for the cleavage of the target DNA and the insertion of the element. Most transposases perform a scattered cleavage of the target DNA generating, after the repair of the insertion site, a characteristic target site duplication (TSD). The size, and sometimes the sequence, of these direct repeats, together with sequence of the TIRs and the structure of the transposase, has been used to classify class II elements in nine different superfamilies: *CACTA*, *Hobo/Activator/Tam3* (*hAT*), *Merlin/IS1016*, *Mutator*, *P-element*, *PIF/Harbinger*, *PiggyBac*, *Tc1/Mariner*, and *Transib* [44–46]. In plants, only elements belonging to the *CACTA*, *hAT*, *Mutator*, *PIF/Harbinger*, and *Tc1/Mariner* superfamilies have been described to date.

MITEs are a particular type of defective class II transposons characterized by their small size and high copy number. Some MITEs show extensive sequence similarities to potential autonomous Class II elements, and it has been shown that the related transposase can bind the TIRs of MITEs [47]. This suggests that, as for the other defective Class II transposons, MITEs could be mobilized by transposases of the elements from which they probably derive [48]. Nevertheless, in other cases no related transposase that could account for the mobilization of the MITE is present in the genome [49, 50]. For this reason, it has been proposed that MITEs could also be mobilized by distantly-related transposases. In support of this hypothesis, phylogenetical and biochemical studies in rice have shown a clear relationship between *Stowaway* MITEs and distantly related *Mariner*-like transposases [51, 52].

Consequences of Transposition: Mutations and More

The mutagenic nature of class II transposons was one of the first characteristics to be associated with these genetic elements but, as other TEs, they can also be at the origin of new genetic variants that can be selected during evolution to ensure new cellular functions. A key difference between Class I and Class II elements is that the insertion of the latter usually produces unstable mutations that can be reverted by the excision of the element. Nevertheless, transposon excision and repair of the empty site usually results in small insertions, deletions or point mutations, known as excision footprints, which can modify the expression or the coding capacity of the mutated gene. This is probably the major source of variability generated by Class II transposons. In some particular cases, transposition can also lead to chromosome breakage or stable chromosomal rearrangements, and it has been recently proposed that these phenomena may have had an important impact on genome evolution [53].

Contrarily to most Class II elements, MITEs can be present at a very high copy number in plant genomes and are frequently polymorphic between individuals of the same species [45, 48]. There are many examples of MITEs being inserted within or close to genes [45, 48] and bioinformatic analyses showed that in some cases the elements close to genes may have been selected during evolution [54]. Nevertheless, in most cases the phenotypic effect of the MITE insertion has not been analyzed in detail.

Transposons are also frequently found closely associated to genes, and their epigenetic silencing can affect the expression of neighboring genes. This is the case of a *Mutator*-like element (*MULE*) inserted within the first intron of the *Arabidopsis FLC* gene, a gene involved in the control of flowering in this plant. It has been recently shown that this *MULE* is targeted by siRNA which directs Histone H3-K9 methylation to the TE region and as a consequence, the *FLC* expression is reduced [55]. The presence of this *MULE* insertion in some ecotypes, such as *Landsberg erecta*, could explain their early flowering phenotype.

Domesticated Transposases as Sources of New Proteins and Protein Domains

Apart from their ability to modify gene expression transposons can be a source of new cellular genes. Transposases are specific DNA-binding proteins containing a catalytic domain responsible for the DNA cleavage and strand transfer reactions that allow transposons to excise from the donor DNA and reinsert into the target DNA. Both the DNA binding and the catalytic activity of transposases are also present in cellular genes that, in some cases, seem to be derived from ancient transposases. This is the case of the human *CENP-B*, a protein that binds the human centromere repeat and is essential for centromere integrity and is presumably derived from a *Mariner*-like transposon, and the *RAG1* protein, derived from a *Transib* transposon, that ensures the V(D)J recombination essential for generating antibody diversity in jaw vertebrates (see [56] for a recent review). In plants, examples of domestication have been described for different families of DNA transposons. A clear example are the *Arabidopsis FAR1* and *FHY3* genes which encode transcription factors involved in the transduction of phytochrome A-mediated signals in plants and whose mutation leads to light-related phenotypes [57]. Both *FAR1* and *FHY3* proteins are closely related to transposases of the *Mutator* family, but none of them is flanked by the typical *Mutator* long TIRs or TSDs, which indicates that *FAR1* and *FHY3* are stable and no longer able to transpose [57]. All these results suggest that both *FAR1* and *FHY3* are transposon-derived cellular genes [57]. In silico searches for other *MULE*s lacking TIRs and TSDs, and evolving under selective pressure (indicative of a potential cellular function), have led to the characterization of the *MUSTANG* family of domesticated transposase genes in *Arabidopsis* and rice [58]. Similarly, the *gary* gene family in cereal grasses was found to be related to the *hAT* family of transposases and to evolve under selective pressure [59]. In this case, *gary* copies have not only lost the TIRs and TSDs, but the *gary* protein does not present anymore the key amino acids required for transposase enzymatic activity [59]. *Gary* seems thus to be a domesticated *hAT*-like transposase that has acquired a yet unknown cellular function.

A functional search for proteins binding the Kubox1 motif present in the upstream region of the *Arabidopsis Ku70* DNA repair gene led to the identification of *DAYSLEEPER*, a gene sharing characteristics of *hAT* transposases and lacking TIRs or TSDs [60]. *DAYSLEEPER* is essential for normal plant growth and its overexpression alters the expression of many genes. Bundock and Hooykaas [60] suggested that *DAYSLEEPER* could be a transcription factor acting through the Kubox1-like boxes that are present in many plant genes.

The examples reviewed above show that domesticated transposases seem to have maintained their ability to specifically bind DNA. Class II transposons can thus have been a source of specific DNA binding domains that, combined with other protein domains, could fulfil new cellular functions, such as transcriptional regulation. But transposase binding sites are found within transposons, and thus the domesticated transposases should have modified their DNA binding specificities to perform the new cellular functions. Alternatively, transposase binding sites could also be present in the cellular locations where the domesticated transposase should function, for example in the promoter regions of a subset of genes to be regulated by the new transposase-transcription factor [61]. Interestingly, MITEs closely associated to genes seem to have been selected during evolution [54], and are frequently found within promoter or terminator regions [48]. A transposon/MITE couple could thus serve to evolve a new transcription factor and distribute throughout the genome its binding sites, in a process that could be seen as the domestication of both the DNA sequences and the encoded proteins of a transposon family. Indeed, this co-domestication seems to be at the origin of the V(D)J recombination system present in jaw vertebrates, as both the *RAG1* core protein and the V(D)J recombination signal sequences seem to be derived from a *Transib* transposon [62].

Transduction of Gene Fragments by Transposons

Pack-MULEs are rice *Mutator*-related elements that have captured genes or gene fragments [63]. Although most of the captured genes seem to be truncated and are not evolving under selective pressure [64], *Pack-MULEs* might have an important impact on host genomes by shuffling protein domains and contributing to the birth of new genes. A non-autonomous element of the *CACTA* family carrying different host gene fragments has also been recently described in soybean [65], suggesting that the ability to capture genes could be a general characteristic of Class II transposons.

Class II Elements: The 'Strange Ones', Helitrons and Foldbacks

Computer analysis of repetitive DNA sequences of *Arabidopsis*, rice and *Caenorhabditis elegans*, revealed the presence of a new family of transposable elements that was called Helitrons [66]. Hypothetical functional Helitrons, reconstituted in silico, are more than 10 kbp long and encode a *HEL* protein composed of the rolling-circle replication initiator and DNA helicase domains necessary for transposition. They also contain a replication protein A (*RPA*)-like with putative single-stranded DNA-binding activity. Based on these characteristics, and in analogy with some bacterial transposable elements, a rolling-circle replication transposition mechanism has been postulated [66]. No functional Helitron has been identified yet but several mutations in maize have been shown to result from Helitron insertions [67]. Unlike most other transposons, Helitrons lack terminal repeats and do not duplicate host sequences during the insertion process. The only invariant Helitron sequences are a 5′ TC end and a 3′ CTAG end preceded in most cases by a 10- to 25-base pair palindrome. They integrate in genomes by cutting an AT dinucleotide.

As for *Pack-MULE* transposons, Helitrons can carry and eventually transcribe fragments of one or several genes, captured from the host genome. In a recent study, Helitrons were implicated in a large number of exon shuffling events leading to intraspecies diversity between two well-known maize lines [68]. The fact that sequences arising from the different captured gene pieces of Helitrons can be conjoined in a chimeric transcript suggests that these elements may promote the emergence of new genes. At least in one case, a complete and functional gene was copied and transposed to a new chromosomal position thereby creating a new maize haplotype [69]. This suggests that Helitrons may play a significant role in creating the genetic heterozygosity exploited in hybrid vigor and other genetic phenomena [67].

Long inverted repeat elements, also known as the Foldback transposons, are a group of modular transposable elements, widespread among eukaryotes. They are variable in size and are capable of forming extensive secondary structure due to the presence of two terminal inverted repeats (IVRs) of several hundred to several thousand bases. Some Foldbacks consist entirely of IVRs while in others the IVRs border a central region of variable size. Although most Foldback elements do not have protein-coding capacity, and therefore are non-autonomous, some copies present large open reading frame that could allow for the production of proteins involved in their mobility. The mechanism by which Foldback moves is unknown but it has been suggested that it involved a DNA intermediate [70]. In plants, typical Foldback elements have been found in rice, in *Arabidopsis*, in rye and in Solanaceae, usually in several hundreds of copies per haploid genome [71–74]. Several copies from the *Arabidopsis FARE2* Foldback family could potentially encode up to three proteins, one of which shows limited homology to a transposase [74].

Studies on *Drosophila* have revealed that Foldbacks have the capacity to generate different types of rearrangements of host chromosomal sequences at unusually high frequencies [75]. Genomic regions, sometimes as large as several hundred kilobases, can transpose to new loci when flanked by Foldback elements and regions flanked by Foldbacks can undergo internal deletions or duplications [75]. Also, a notable feature of Foldbacks in *Drosophila* is their ability to undergo high frequency precise excision creating mutational hot spots [76]. Our knowledge on plant Foldbacks is still at an early stage and we do not know yet if they were major actor in shaping plant genome structure, as this was obviously the case for *Drosophila*.

Conclusion

A fascinating picture is emerging where TEs appear as pivotal factors in generating variation and large-scale reprogramming of gene expression. As a consequence, a more integrate view taking into account the role of TEs in formatting the structure and regulatory architecture of genomes is now needed to better understand genome function and phenotypic diversity. Higher plants are usually very rich in TEs, yet large scale studies to evaluate the importance of TE contribution to the regulatory diversity of plant genes are scarce. A lot more remains to be discovered and examples of TEs having key impacts on genome organization or regulating key cellular functions are likely to accumulate in the near future.

Note Added in Proof

After the submission of our manuscript, the mobilisation of plant MITEs by class II related transposases was demonstrated (Yang G, Zhang F, Hancock CN, Wessler SR: Transposition of the rice miniature inverted repeat transposable element mPing in *Arabidopsis thaliana*. Proc Natc Acad Sci USA 2007;104:10962–10967).

References

1 L'Homme Y, Seguin A, Tremblay FM: Different classes of retrotransposons in coniferous spruce species. Genome 2000;43:1084–1089.

2 Vitte C, Bennetzen JL: Analysis of retrotransposon structural diversity uncovers properties and propensities in angiosperm genome evolution. Proc Natl Acad Sci USA 2006;103:17638–17643.

3 Witte CP, Le QH, Bureau T, Kumar A: Terminal-repeat retrotransposons in miniature (TRIM) are involved in restructuring plant genomes. Proc Natl Acad Sci USA 2001;98:13778–13783.

4 Kalendar R, Vicient CM, Peleg O, Anamthawat-Jonsson K, Bolshoy A, Schulman AH: Large retrotransposon derivatives: abundant, conserved but nonautonomous retroelements of barley and related genomes. Genetics 2004;166:1437–1450.

5 Vitte C, Panaud O: LTR retrotransposons and flowering plant genome size: emergence of the increase/decrease model. Cytogenet Genome Res 2005;110:91–107.

6 Yan H, Ito H, Nobuta K, Ouyang S, Jin W, et al: Genomic and genetic characterization of rice Cen3 reveals extensive transcription and evolutionary implications of a complex centromere. Plant Cell 2006;18:2123–2133.

7 SanMiguel P, Tikhonov A, Jin YK, Motchoulskaia N, Zakharov D, et al: Nested retrotransposons in the intergenic regions of the maize genome. Science 1996;274:765–768.

8 Bennetzen JL, Ma J, Devos KM: Mechanisms of recent genome size variation in flowering plants. Ann Bot (Lond) 2005;95:127–132.

9 Chaparro C, Guyot R, Zuccolo A, Piegu B, Panaud O: RetrOryza: a database of the rice LTR-retrotransposons. Nucleic Acids Res 2007;35:D66–D70.

10 Hawkins JS, Kim H, Nason JD, Wing RA, Wendel JF: Differential lineage-specific amplification of transposable elements is responsible for genome size variation in *Gossypium*. Genome Res 2006;16:1252–1261.

11 Piegu B, Guyot R, Picault N, Roulin A, Saniyal A, et al: Doubling genome size without polyploidization: dynamics of retrotransposition-driven genomic expansions in *Oryza australiensis*, a wild relative of rice. Genome Res 2006;16:1262–1269.

12 SanMiguel P, Gaut BS, Tikhonov A, Nakajima Y, Bennetzen JL: The paleontology of intergene retrotransposons of maize. Nat Genet 1998;20:43–45.

13 Vitte C, Ishii T, Lamy F, Brar D, Panaud O: Genomic paleontology provides evidence for two distinct origins of Asian rice (*Oryza sativa* L.). Mol Genet Genomics 2004;272:504–511.

14 Ma J, Bennetzen JL: Rapid recent growth and divergence of rice nuclear genomes. Proc Natl Acad Sci USA 2004;101:12404–12410.

15 Ma J, SanMiguel P, Lai J, Messing J, Bennetzen JL: DNA rearrangement in orthologous orp regions of the maize, rice and sorghum genomes. Genetics 2005;170:1209–1220.

16 Kashkush K, Feldman M, Levy AA: Transcriptional activation of retrotransposons alters the expression of adjacent genes in wheat. Nat Genet 2003;33:102–106.

17 Grandbastien MA, Audeon C, Bonnivard E, Casacuberta JM, Chalhoub B, et al: Stress activation and genomic impact of Tnt1 retrotransposons in Solanaceae. Cytogenet Genome Res 2005;110: 229–241.

18 Sugimoto K, Takeda S, Hirochika H: Transcriptional activation mediated by binding of a plant GATA-type zinc finger protein AGP1 to the AG-motif (AGATCCAA) of the wound-inducible Myb gene NtMyb2. Plant J 2003;36:550–564.

19 Moran J, Gilbert N: Mammalian LINE-1 Retrotransposons and related elements; in Craig N, Craigie R, Gellert M, Lambowitz AM (eds): Mobile DNA II. ASM Press, Washington DC, 2002, pp 836–869.

20 Schwarz-Sommer Z, Leclercq L, Gobel E, Saedler H: Cin4, an insert altering the structure of the *A1* gene in *Zea mays*, exhibits properties of nonviral retrotransposons. EMBO J 1987;6:3873–3880.

21 Schmidt T: LINEs, SINEs and repetitive DNA: non-LTR retrotransposons in plant genomes. Plant Mol Biol 1999;40:903–910.

22 Leeton PR, Smyth DR: An abundant LINE-like element amplified in the genome of *Lilium speciosum*. Mol Gen Genet 1993;237:97–104.

23 Komatsu M, Shimamoto K, Kyozuka J: Two-step regulation and continuous retrotransposition of the rice LINE-type retrotransposon Karma. Plant Cell 2003;15:1934–1944.

24 Yamashita H, Tahara M: A LINE-type retrotransposon active in meristem stem cells causes heritable transpositions in the sweet potato genome. Plant Mol Biol 2006;61:79–94.

25 Han JS, Szak ST, Boeke JD: Transcriptional disruption by the L1 retrotransposon and implications for mammalian transcriptomes. Nature 2004;429:268–274.

26 Casacuberta E, Pardue ML: HeT-A and TART, two *Drosophila* retrotransposons with a bona fide role in chromosome structure for more than 60 million years. Cytogenet Genome Res 2005;110: 152–159.

27 Dewannieux M, Esnault C, Heidmann T: LINE-mediated retrotransposition of marked Alu sequences. Nat Genet 2003;35:41–48.

28 Kajikawa M, Okada N: LINEs mobilize SINEs in the eel through a shared 3′ sequence. Cell 2002;111:433–444.

29 Kramerov DA, Vassetzky NS: Short retroposons in eukaryotic genomes. Int Rev Cytol 2005;247:165–221.

30 Borodulina OR, Kramerov DA: Wide distribution of short interspersed elements among eukaryotic genomes. FEBS Lett 1999;457:409–413.
31 Deragon JM, Landry BS, Pelissier T, Tutois S, Tourmente S, Picard G: An analysis of retroposition in plants based on a family of SINEs from *Brassica napus*. J Mol Evol 1994;39:378–386.
32 Umeda M, Ohtsubo H, Ohtsubo E: Diversification of the rice *Waxy* gene by insertion of mobile DNA elements into introns. Jpn J Genet 1991;66:569–586.
33 Yasui Y, Nasuda S, Matsuoka Y, Kawahara T: The Au family, a novel short interspersed element (SINE) from *Aegilops umbellulata*. Theor Appl Genet 2001;102:463–470.
34 Yoshioka Y, Matsumoto S, Kojima S, Ohshima K, Okada N, Machida Y: Molecular characterization of a short interspersed repetitive element from tobacco that exhibits sequence homology to specific tRNAs. Proc Natl Acad Sci USA 1993;90:6562–6566.
35 Shedlock AM, Okada N: SINE insertions: powerful tools for molecular systematics. Bioessays 2000;22:148–160.
36 Cheng C, Motohashi R, Tsuchimoto S, Fukuta Y, Ohtsubo H, Ohtsubo E: Polyphyletic origin of cultivated rice: based on the interspersion pattern of SINEs. Mol Biol Evol 2003;20:67–75.
37 Tatout C, Warwick SI, Lenoir A, Deragon JM: SINE insertion as clade markers for wild Cruciferae species. Mol Biol Evol 1999;16:1614–1621.
38 Motohashi R, Mochizuki K, Ohtsubo H, Ohtsubo E: Structure and distribution of p-SINE1 members in rice genomes. Theor Appl Genet 1997;95:359–368.
39 Jurka J, Kapitonov VV, Pavlicek A, Klonowski P, Kohany O, Walichiewicz J: Repbase Update, a database of eukaryotic repetitive elements. Cytogenet Genome Res 2005;110:462–467.
40 Pozueta-Romero J, Houlne G, Schantz R: Identification of a short interspersed repetitive element in partially spliced transcripts of the bell pepper (*Capsicum annuum*) PAP gene: new evolutionary and regulatory aspects on plant tRNA-related SINEs. Gene 1998;214:51–58.
41 Lenoir A, Cournoyer B, Warwick S, Picard G, Deragon JM: Evolution of SINE S1 retroposons in Cruciferae plant species. Mol Biol Evol 1997;14:934–941.
42 Deragon JM, Zhang X: Short Interspersed Elements (SINEs) in plants: Origin, classification and use as phylogenetic markers. Syst Biol 2006;55:949–956.
43 Sun FJ, Fleurdepine S, Bousquet-Antonelli C, Caetano-Anolles G, Deragon JM: Common evolutionary trends for SINE RNA structures. Trends Genet 2007;23:26–33.
44 Feschotte C: Merlin, a new superfamily of DNA transposons identified in diverse animal genomes and related to bacterial IS1016 insertion sequences. Mol Biol Evol 2004;21:1769–1780.
45 Feschotte C, Jiang N, Wessler SR: Plant transposable elements: where genetics meets genomics. Nat Rev Genet 2002;3:329–341.
46 Kapitonov VV, Jurka J: Molecular paleontology of transposable elements in the *Drosophila melanogaster* genome. Proc Natl Acad Sci USA 2003;100:6569–6574.
47 Loot C, Santiago N, Sanz A, Casacuberta JM: The proteins encoded by the pogo-like Lemi1 element bind the TIRs and subterminal repeated motifs of the *Arabidopsis* Emigrant MITE: consequences for the transposition mechanism of MITEs. Nucleic Acids Res 2006;34: 5238–5246.
48 Casacuberta JM, Santiago N: Plant LTR-retrotransposons and MITEs: control of transposition and impact on the evolution of plant genes and genomes. Gene 2003;311:1–11.
49 Jiang N, Bao Z, Zhang X, Hirochika H, Eddy SR, McCouch SR, Wessler SR: An active DNA transposon family in rice. Nature 2003;421:163–167.
50 Kikuchi K, Terauchi K, Wada M, Hirano HY: The plant MITE mPing is mobilized in anther culture. Nature 2003;421:167–170.
51 Feschotte C, Osterlund MT, Peeler R, Wessler SR: DNA-binding specificity of rice mariner-like transposases and interactions with Stowaway MITEs. Nucleic Acids Res 2005;33:2153–2165.
52 Feschotte C, Swamy L, Wessler SR: Genome-wide analysis of mariner-like transposable elements in rice reveals complex relationships with stowaway miniature inverted repeat transposable elements (MITEs). Genetics 2003;163:747–758.
53 Zhang J, Peterson T: A segmental deletion series generated by sister-chromatid transposition of Ac transposable elements in maize. Genetics 2005;171:333–344.
54 Santiago N, Herraiz C, Goni JR, Messeguer X, Casacuberta JM: Genome-wide analysis of the Emigrant family of MITEs of *Arabidopsis thaliana*. Mol Biol Evol 2002;19:2285–2293.

55 Liu J, He Y, Amasino R, Chen X: siRNAs targeting an intronic transposon in the regulation of natural flowering behavior in *Arabidopsis*. Genes Dev 2004;18:2873–2878.
56 Volff JN: Turning junk into gold: domestication of transposable elements and the creation of new genes in eukaryotes. Bioessays 2006;28:913–922.
57 Hudson ME, Lisch DR, Quail PH: The *FHY3* and *FAR1* genes encode transposase-related proteins involved in regulation of gene expression by the phytochrome A-signaling pathway. Plant J 2003;34:453–471.
58 Cowan RK, Hoen DR, Schoen DJ, Bureau TE: MUSTANG is a novel family of domesticated transposase genes found in diverse angiosperms. Mol Biol Evol 2005;22:2084–2089.
59 Muehlbauer GJ, Bhau BS, Syed NH, Heinen S, Cho S, et al: A hAT superfamily transposase recruited by the cereal grass genome. Mol Genet Genomics 2006;275:553–563.
60 Bundock P, Hooykaas P: An *Arabidopsis* hAT-like transposase is essential for plant development. Nature 2005;436:282–284.
61 Cordaux R, Udit S, Batzer MA, Feschotte C: Birth of a chimeric primate gene by capture of the transposase gene from a mobile element. Proc Natl Acad Sci USA 2006;103:8101–8106.
62 Kapitonov VV, Jurka J: RAG1 core and V(D)J recombination signal sequences were derived from Transib transposons. PLoS Biol 2005;3:e181.
63 Jiang N, Bao Z, Zhang X, Eddy SR, Wessler SR: Pack-MULE transposable elements mediate gene evolution in plants. Nature 2004;431:569–573.
64 Juretic N, Hoen DR, Huynh ML, Harrison PM, Bureau TE: The evolutionary fate of MULE-mediated duplications of host gene fragments in rice. Genome Res 2005;15:1292–1297.
65 Zabala G, Vodkin LO: The wp mutation of *Glycine max* carries a gene-fragment-rich transposon of the CACTA superfamily. Plant Cell 2005;17:2619–2632.
66 Kapitonov VV, Jurka J: Rolling-circle transposons in eukaryotes. Proc Natl Acad Sci USA 2001;98:8714–8719.
67 Lal SK, Hannah LC: Plant genomes: massive changes of the maize genome are caused by Helitrons. Heredity 2005;95:421–422.
68 Morgante M, Brunner S, Pea G, Fengler K, Zuccolo A, Rafalski A: Gene duplication and exon shuffling by helitron-like transposons generate intraspecies diversity in maize. Nat Genet 2005;37:997–1002.
69 Xu JH, Messing J: Maize haplotype with a helitron-amplified cytidine deaminase gene copy. BMC Genet 2006;7:52.
70 Potter SS: DNA sequence of a foldback transposable element in *Drosophila*. Nature 1982;297: 201–204.
71 Alves E, Ballesteros I, Linacero R, Vazquez AM: RYS1, a foldback transposon, is activated by tissue culture and shows preferential insertion points into the rye genome. Theor Appl Genet 2005;111:431–436.
72 Daskalova SM, Scott NW, Elliott MC: Folbos, a new foldback element in rice. Genes Genet Syst 2005;80:141–145.
73 Rebatchouk D, Narita JO: Foldback transposable elements in plants. Plant Mol Biol 1997;34:831–835.
74 Windsor AJ, Waddell CS: FARE, a new family of foldback transposons in *Arabidopsis*. Genetics 2000;156:1983–1995.
75 Collins M, Rubin GM: Structure of chromosomal rearrangements induced by the FB transposable element in *Drosophila*. Nature 1984;308:323–327.
76 Collins M, Rubin GM: High-frequency precise excision of the *Drosophila* foldback transposable element. Nature 1983;303:259–260.

Jean-Marc Deragon
LGDP, UMR CNRS-IRD-UP 5096
Université de Perpignan, 52, Av. Paul Alduy
FR-66860 Perpignan cedex (France)
Tel. +33 4 68 66 17 73, Fax +33 4 68 66 84 99, E-Mail jean-marc.deragon@univ-perp.fr

Volff J-N (ed): Plant Genomes.
Genome Dyn. Basel, Karger, 2008, vol 4, pp 83–94

Plant Sex Chromosomes

D. Charlesworth

Institute of Evolutionary Biology, School of Biological Sciences, University of Edinburgh, Ashworth Lab. King's Buildings, Edinburgh, UK

Abstract

Dioecious species are known in plants and, as in many animals, some have distinguishable sex chromosomes. Genetic maps have identified sex-determining regions in several plants, and mapped male-specific Y (MSY) regions of the chromosome in which crossing over and genetic recombination do not occur, allowing sequence divergence between the X and Y. Divergence values of the few X-Y gene pairs so far available show that recombination between different genes of *Silene latifolia* stopped at different times. Once recombination stops, MSY genome regions are predicted to accumulate repetitive sequences, including transposable elements, resulting in low gene density. This has been documented in papaya but not yet in other plants. Y-linked genes should also accumulate deleterious mutations, eventually being lost as dosage compensation evolves. The few available data suggest that many plant MSY genes are functional, perhaps because genes required for male gametophyte functions degenerate slowly. Detailed studies of sex-linked genes are needed to test for deleterious substitutions in Y genes, and to date the origins of plant sex chromosomes.

The term 'sex chromosomes' is used when a chromosome pair carrying sex determining genes differ in size or shape in the two sexes (heteromorphism) and is largely non-recombining. In mammals and *Drosophila*, males are the heterogametic sex, with X and Y chromosomes [reviewed in 1, 2] and birds have female heterogamety [Z/W sex chromosomes, reviewed in 3]. Y chromosomes in mammals and *Drosophila* species are 'genetically degenerated' – carrying few genes compared with the X. Degeneration probably results from loss of recombination, as it also occurs in non-sex chromosome systems, including fungal incompatibility gene regions [4], which do not specify different sexes (i.e. different gamete sizes).

Heteromorphic sex chromosomes are present in some dioecious plants. Most have male heterogamety, as in *Silene latifolia* and *Rumex* species, two plants that have become 'model systems' for sex chromosome studies [5, 6]. In some dioecious plants, males are heterozygous for a sex-determining region in which recombination is suppressed, but the sex chromosomes are not heteromorphic. There is a male-specific Y-like (MSY) region isolated from the homologous X region, and differentiated in sequence, so the chromosome may be called a Y, although it is not cytologically recognisable. Heteromorphic sex chromosomes are also known in bryophytes such as *Marchantia polymorpha* [7]. In these haploid plants, Y and X chromosomes refer to the chromosomes found in male and female gametophytes, respectively. Here, I shall refer only to X/Y systems, but most of the concepts discussed apply to female heterogamety, of which some cases are known in plants [5].

Understanding sex chromosome evolution is an important part of understanding low recombination genome regions. Sex chromosomes that evolved in the past few million years can shed light on the initial steps in the evolution of the differences between X and Y chromosomes. Thus plant sex chromosomes have attracted attention, since their scattered distribution suggests that they have arisen at least several times in different plant taxa, often recently (see below). Specifically, it is of interest to try to establish the amounts of times needed for the evolution of the special characteristics of Y chromosomes, including genetic degeneration, accumulation of repetitive DNA sequences, heterochromatinization, chromosome rearrangements (inversions, duplications and deletions), possible evolution of a bias towards male functions, and of dosage compensation, in which expression levels of X-linked genes are adjusted in the two sexes to compensate for the difference between two X copies in females and only one in males. Each of these topics will be briefly reviewed here, focusing on recent advances.

Plant Sex Chromosomes and Sex-Determining Regions

To date, no plant sex-determining locus has yet been identified, but in several dioecious plants, sex-determining regions have been identified in genetic maps, usually using anonymous markers (table 1). In some species, multiple completely linked markers suggest that recombination is suppressed across part of the chromosome, while the rest is 'pseudo-autosomal' like the recombining tips of the human X and Y. Such map data over-estimate the physical size of the non-recombining region relative to other genome regions, because markers that co-segregate in a mapping population may nevertheless recombine at a low frequency, and also because markers are most abundant in non-recombining regions (due to their divergence and rearrangement, see below). Yet, in some plants, only partially linked markers, and no completely linked ones, are so far

Table 1. Genetic maps of angiosperm sex chromosomes. Species are listed starting with those whose sex chromosomes are mostly pseudo-autosomal (PSA), indicated with black (possibly the youngest plant sex chromosome systems), and ending with species whose Y chromosomes are mostly male-specific (indicated in grey), and whose sex chromosomes are possibly older. Almost all markers are anonymous, including AFLP and RAPD, but use of genic or microsatellite markers are indicated by footnotes.

Species	Sex chromosomes	Proportion of PSA markers	References
Species with mostly pseudo-autosomal sex chromosomes			
Asparagus officinalis		All 33 markers on the chromosome	[54]
Actinidia chinensis	Probably X/Y	Most	[55, 56]
Carica papaya	Males heterozygous	Most	[26]
Vitis vinifera	X/Y	Most	[57]
Spinacia oleracea	Uncertain (males heterozygous)	Most	[58]
Species with mostly male-specific Y chromosomes			
Cannabis sativa	Uncertain (males heterozygous)	Several[a]	[59, 60]
Silene latifolia	X/Y (pairing only at tip)	0 out of 6 markers[b]; Many	[11, 38, 61]
Humulus lupulus	X/Y (pairing only at tip)	Unknown	[62]

[a]Some microsatellite markers; [b]Genic markers.

known. These sex chromosomes probably have very small non-recombining regions, and they are often considered to be the youngest plant sex chromosomes, though this is not necessarily correct. We return below to estimating their ages.

An important type of evidence to supplement genetic maps is to test for associations between marker alleles and sex. Complete association with one sex in multiple natural populations suggests that recombination is very rare [e.g. 8], and can be easier to demonstrate than close linkage within a mapping family (which requires that the family includes suitable marker variants in several loci). Different allele frequencies between the sexes should allow pseudo-autosomal loci to be detected [9], but so far such loci have not been studied in plants with large MSY regions (classical sex chromosomes); all markers so far mapped to such regions are anonymous (table 1).

Why Is Recombination Suppressed in Sex-Determining Regions?

Recombination suppression preserves the different combinations of genes appropriate for males and females. Two distinct situations select for reduced recombination. In the initial stages of sex chromosome evolution from an hermaphroditic or monoecious ancestor, females probably arise by male sterility mutations at a nuclear male function gene, and invade the population, which becomes gynodioecious (with both female and functionally hermaphrodite plants). Alleles increasing the hermaphrodites' male fertility are then favoured, even if female fertility is reduced [10]. Selection for restricted recombination arises because recombinants involving these two sex-determining genes are disfavoured. If the mutations are expressed in both sexes, 'trade-offs' between increased male and decreased female fertility select against the appearance in females of these suppressors of female functions. If the mutant effect were sex-limited, i.e. not expressed in female development, any decrease in female fertility would affect only hermaphrodites, and the mutations could spread provided only that they sufficiently increase male fertility to outweigh the lost female function; the recombination rate with the male-sterility locus would be irrelevant. Thus, in the transition from gynodioecy to dioecy, mutations not closely linked to the male-sterility locus often cannot invade. This predicts that the male- and female-suppressing mutations involved must arise on the same chromosome pair, and modifiers reducing recombination between these loci will also be favoured [10]. The non-dioecious plant *Silene vulgaris* indeed has a chromosome with the same gene content as the X of the dioecious *S. latifolia* [11].

Once such a proto-Y chromosome is established, it may accumulate further sexually antagonistic genes, increasing male functions but reducing female fertility, since the evolution of females alters the optimal balance of resource allocation of the hermaphrodites. Genes on the evolving Y affecting flower functions may thus evolve to specialize in male functions [10, 12]. These new evolutionary changes select for further recombination suppression between the Y and X. Thus recombination suppression is likely to evolve in several stages (see below).

Primary sex-determination need not involve two genes. In some insects, established systems have been replaced by a single gene [e.g. 13]. This possible alternative to de novo evolution of separate sexes should be considered as a possibility for species whose sex-determining region is physically small, particularly when related species are also dioecious, e.g. in fish and perhaps some plants, though no case is yet known. Sexually antagonistic genes could accumulate in the region and select for reduced recombination in such systems.

The Fate of Plant Sex-Linked Genes and Sex-Determining Regions

Although identifying the primary sex-determining genes may be of interest, the argument above suggests that these are likely often to be ordinary genes whose mutations can cause male and female sterility. A question of equal interest is how Y chromosomes evolve, including the time-scale of the changes from an ordinary chromosome to a genetically degenerate one.

Finding multiple markers confined to males is only the first step in establishing that a proto-Y chromosome is evolving. If a region of suppressed recombination has evolved around a species' sex-determining genes, its loss of recombination will have several effects on the evolution of genes in the region. Without recombination, genetic hitch-hiking processes [reviewed in 14] occur that reduce the effective population size of Y-linked genes below the values expected from the relative numbers of copies of Y-linked, X-linked and autosomal alleles [15]. Hitch-hiking will occur as advantageous mutations spread through the species' Y chromosomes [16, 17]; all Y chromosomes will have a common ancestor at the time of the most recent such event (a 'selective sweep'). Hitch-hiking also occurs as natural selection removes deleterious mutations from populations, lowering the ability of selective forces to maintain adaptation in the face of deleterious mutations [14] and to fix adaptive changes in the species [18].

Low effective population size of Y-linked genes allows the accumulation of transposable elements and other repetitive sequences. Insertion of such sequences is often deleterious, and they accumulate preferentially in genome regions with low recombination rates [19], including the evolving *Drosophila miranda* neo-Y chromosome [17], and plant centromere regions [e.g. 20, 21]. The larger size of many plant Y chromosomes than their Xs may be due to such accumulation [22]. Although repetitive sequences have been detected in dioecious plants [e.g. 23, 24], only a few are Y-specific; these include some of the markers used for the mapping studies in table 1, and others include a satellite sequence in *Rumex* species [6], retrotransposon sequences in *S. latifolia* [25], a LINE-like retrotransposon tandemly repeated at the end of the long arm of the *Cannabis sativa* Y [22], and repeated sequences in *Marchantia polymorpha* [7].

Until related species are compared, it is unknown whether accumulation has occurred on the Y chromosomes (some plant genomes have high contents of transposable elements and other repetitive sequences on all chromosomes that have been sampled). In papaya, it seems likely that higher Y repetitive content [26] is due to the predicted accumulation, rather than a reduction on the X. The abundance of Y-linked markers in dioecious plants is also consistent with this. Accumulation of repetitive sequences makes Y sequences increasingly different from the X, creating numerous Y-linked markers if the region lacking recombination is extensive. It will be important in the future to use outgroup species to test whether the Y or the X has changed, and distinguish between accumulation of repeat sequences on the Y, versus a general increase in both sex chromosomes, or even a genome-wide increase. Once repetitive sequences are present, generating inter-genic spaces or long introns with inserted sequences not present in X genes, insertion of further sequences is more likely to occur within these regions without damaging functional sites, as appears to have occurred in maize [27]. Thus the rate of accumulation will accelerate. Examples of fast accumulation are known in plant genomes [28], but the time-scale of accumulation is not yet known for plant Y chromosomes.

Are Plant Y Chromosomes Genetically Degenerated?

A lower Y gene density may reflect accumulation of repetitive sequences, and does not necessarily imply loss of Y-linked genes. A low gene density, or an accumulation of repetitive sequences might affect purely inter-genic sequences and/or introns, potentially leaving gene functions intact. It should therefore not be termed 'degeneration' without explicit testing for loss of genes, since all the genes present on the ancestral chromosome may still be present. Repetitive sequences may, however, be involved in degeneration by increasing the rate of chromosome rearrangement mutations, potentially leading to deletion of genes.

Genetic degeneration (i.e. loss of genes or deterioration of gene functions), has probably occurred in some plant sex chromosome systems. In some species an X chromosome is required for viability, and YY genotypes are often absent in progeny of crosses between functionally hermaphrodite plants with deletions of part of the Y [5]. Haploid *Silene latifolia* plants regenerated from pollen almost all have the X, not the Y [29], hermaphrodite plants' Y-bearing pollen sires only small fractions of seeds [30], and Y-bearing ovules appear to be inviable [31]. However, a single lethal Y-linked mutation could be responsible, potentially just in the strain studied, so further tests are needed to learn whether loci are degenerating.

It is not yet known what fraction of any dioecious plant's genes are degenerated, nor what form degeneration takes. These questions require study of Y-linked genes, not just anonymous markers, and comparisons with X-linked homologues. Reduced effective population sizes of Y-linked loci should reduce adaptation compared with their X-linked homologues, and deleterious variants such as amino acid replacements that reduce the function of the protein encoded may fix, as has happened in *Drosophila* [17]. Degeneration can thus be detected before genes are lost, and one plant example of a non-functional Y gene copy has been found [32].

Finding sex-linked genes in non-model species is, however, not easy. Little is known about the genomes of most dioecious plants. Given a genetic map, or using in situ hybridization, the chromosome carrying the sex-determining genes can sometimes be identified, and BAC clones from the chromosome can be sequenced. In Asparagus [33] or *Marchantia* [7], the high repetitive sequence content means that genes are rarely found [26, 34]. Random amplification of sequences from isolated X or Y chromosomes has similar drawbacks.

Another approach, so far used only in *Silene latifolia*, uses ESTs to obtain sequences for primer design and genic marker development to find sex-linked loci [32, 35–39]. The estimated sequence divergence between the few gene pairs so far known with both X- and Y-linked copies show clearly that, at least in *S. latifolia*, recombination suppression has occurred in several stages, as predicted above. Divergence for synonymous or intron sites (under slight selective constraint), depends mainly on the time since the two sequences stopped recombining and started diverging. In *S. latifolia*, there is a large range of divergence values, like those found between the mammalian X and Y chromosomes ('evolutionary strata', see [40–42]), although the evolutionary time-scale is much shorter for *Silene* [11, 38]. The highest silent or synonymous site divergence estimated for *S. latifolia* X-Y gene pairs is around 20%, suggesting that recombination in these regions ceased about 10 million years ago. As in the mammalian X chromosome genetic map, the oldest sex-linked genes (most diverged pairs) are furthest from the pseudo-autosomal region [38].

Most *S. latifolia* sex-linked genes so far identified have X and Y copies, and the Y coding sequences give little evidence of degeneration, based on testing for excess deleterious mutations in the Y copies, compared with the X, after correcting for the possibility of mutation rate differences [43]. One gene, *SlY3*, has substituted an unexpectedly large number of amino acids since its split from the X lineage, and also appears to have low expression [38]. Since in angiosperms a high proportion of genes are expressed in the haploid pollen stage [44], many are probably essential for both X- or Y-bearing pollen; thus degeneration may affect fewer genes than in animals (whose male gametes often do not express their genes).

The fundamental cause of genetic degeneration is probably failure to recombine with the homologous X chromosome region [14]. The heterozygous state of Y chromosomes in diploids, allows non-expression of deleterious mutations unless their effects show some dominance [45]. Indeed repetitive sequences also accumulate in non-recombining genome regions in haploids, e.g. in incompatibility gene regions of some fungi [4]. Study of the X chromosome of the haploid bryophyte plant *M. polymorpha* will thus be very interesting. Because the

X occurs only in female gametophytes, and the Y in male ones, both are similarly non-recombining, and they should degenerate similarly (and no dosage compensation needs to evolve). Degeneration may, however, be slow, as mutations are not sheltered in heterozygotes; X- and Y-linked genes essential for the gametophyte stage should be maintained (like angiosperm genes with essential pollen functions and many housekeeping genes of the Y are indeed also present in females [46], but genes involved in one or other sex functions may degenerate. Repetitive sequences have accumulated on the Y [7], but tests for genetic degeneration, and studies of the X, have not yet been done.

The Age of Plant Sex Chromosomes

So far, the only sequence data from plant sex-linked gene pairs are from *S. latifolia*. Other plant sex chromosomes, with smaller non-recombining regions (table 1) may not yet have evolved suppressed recombination, or have only started to do so, and it is often stated that these are young sex chromosomes. Alternatively, the lack of pronounced secondary sex differences in plants (apart from flower characteristics) probably limits the opportunity for sexually antagonistic mutations. Thus multiple steps of recombination reduction may sometimes not occur, even in old sex chromosome systems. To test between these possibilities, plant sex chromosome ages must be independently estimated. Detailed studies of young sex chromosomes should establish the amounts of times needed for the evolution of different characteristics of Y chromosomes, and the time-scales will also help to distinguish the processes responsible. The rates of at least four processes are of interest: (i) accumulation of repetitive sequences; (ii) chromosome rearrangements, which may contribute to recombination suppression and to degenerative processes (for instance by causing gene loss); (iii) reduced gene expression or protein function ('genetic degeneration'), and (iv) the evolution of dosage compensation.

To progress in understanding these processes, sex chromosome genes must be identified and studied in more plant systems. To discover the most diverged gene pairs, which date the sex chromosomes' origin (i.e. the time when X-Y recombination ceased), multiple genes are needed. Finding low divergence genes as in papaya [47] is not convincing evidence for a recent origin, particularly if several genes are ascertained from physically close regions (e.g. from a single BAC clone), which probably stopped recombining at similar times; the full range of divergence values is needed.

More phylogenetic work will also help. It is thought that sex chromosomes evolved since the split of species within the genus *Silene* [48], but the placement of *S. latifolia* among non-dioecious ancestors is based on a phylogeny estimated only from short ITS sequences. Without independent evidence from further nuclear genes, an older history cannot be firmly excluded. *Silene* Y chromosomes are old enough to have increased greatly in size compared with the X, and to have undergone rearrangements [49], but are not constitutively heterochromatic. Even the older *Rumex* Y chromosomes, which have undergone X-autosome fusion, creating a lineage with a neo-Y [50], may not be fully constitutively heterochromatic [51].

Conclusions and Future Prospects

The question remains whether plant Y chromosomes have evolved a distinctively male character; in some plants, there may have been insufficient time, and in some the sex roles may not differ enough. Despite its large size, even the *S. latifolia* Y may retain a similar gene content to that of the X. Many genes may have male gametophyte functions and will not degenerate. Indeed, most Y deletions yield low male frequencies in the offspring, i.e. they greatly lower the chance of Y-bearing pollen grains achieving fertilisation [30]. In addition, genes with male fertility functions may have evolved alleles enhancing maleness (including sexually antagonistic alleles) or have been moved to the Y. Only one case of transposition of a gene to a plant Y is, however so far known, in *S. latifolia* [52]. Many *M. polymorpha* genes found on the Y, but absent in females, have likely male functions [47].

If degeneration has started, plants will be particularly interesting for studies of dosage compensation. It should be possible to test whether individual genes are compensated, or whether the evolution of dosage compensation affects larger genome regions. Plants also offer opportunities to study the time-scale of de novo evolution of dosage compensation. In the animal systems so far studied (particularly neo-sex chromosomes in *Drosophila*, with an autosome translocated to a sex chromosome, becoming restricted to passage through males, and thus ceasing to recombine), there is a pre-existing dosage compensation system, so that the cellular machinery involved does not have to evolve [53]. Thus neo-Y chromosomes might evolve dosage compensation faster than would occur de novo. At present, understanding plant sex chromosomes is at an early stage, but now that it is possible to discover genes on X and Y chromosomes of dioecious species, many of the interesting questions should yield answers.

Acknowledgements

This work was supported by the Natural Environment Research Council of the UK. I thank R. Bergero and V. Kaiser for discussions, comments on the manuscript, and unpublished data.

References

1 Vallender EJ, Lahn BT: How mammalian sex chromosomes acquired their peculiar gene content. Bioessays 2004;26:159–169.
2 Carvalho AB: Origin and evolution of the *Drosophila* Y chromosome. Curr Opin Genet Dev 2002;12:664–668.
3 Ellegren H: Evolution of avian sex chromosomes. Trends Eco Evol 2000;15:188–191.
4 Fraser JA, Diezmann S, Subaran RL, Allen A, Lengeler KB, et al: Convergent evolution of chromosomal sex-determining regions in the animal and fungal kingdoms. PLoS Biol 2004;2: 2243–2255.
5 Westergaard M: The mechanism of sex determination in dioecious plants. Adv Genet 1958;9: 217–281.
6 Shibata F, Hizume M, Kuroki Y: Differentiation and the polymorphic nature of the Y chromosomes revealed by repetitive sequences in the dioecious plant, *Rumex acetosa*. Chromosome Res 2000;8:229–236.
7 Ishizaki K, Shimizu-Ueda Y, Okada S, Yamamoto M, Fujisawa M, et al: Multicopy genes uniquely amplified in the Y chromosome-specific repeats of the liverwort *Marchantia polymorpha*. Nucleic Acids Res 2002;30:4675–4681.
8 Filatov DA, Monéger F, Negrutiu I, Charlesworth D, et al: Evolution of a plant Y-chromosome: variability in a Y-linked gene of *Silene latifolia*. Nature 2000;404:388–390.
9 Marshall AR, Knudsen KL, Allendorf FW: Linkage disequilibrium between the pseudoautosomal PEPB-1 locus and the sex-determining region of chinook salmon. Heredity 2004;93:85–97.
10 Charlesworth B, Charlesworth D: A model for the evolution of dioecy and gynodioecy. Am Nat 1978;112:975–997.
11 Filatov DA: Evolutionary history of *Silene latifolia* sex chromosomes revealed by genetic mapping of four genes. Genetics 2005;170:975–979.
12 Rice WR: The accumulation of sexually antagonistic genes as a selective agent promoting the evolution of reduced recombination between primitive sex-chromosomes. Evolution 1987;41: 911–914.
13 Traut W, Sahara K, Otto TD, Marec F: Molecular differentiation of sex chromosomes probed by comparative genomic hybridization. Chromosoma 1999;108:173–180.
14 Charlesworth B, Charlesworth D: The degeneration of Y chromosomes. Phil Trans R Soc Lond B 2000;355:1563–1572.
15 Laporte V, Charlesworth B: Effective population size and population subdivision in demographically structured populations. Genetics 2002;162:501–519.
16 Rice WR: Genetic hitch-hiking and the evolution of reduced genetic activity of the Y sex chromosome. Genetics 1987;116:161–167.
17 Bachtrog D: A dynamic view of sex chromosome evolution. Curr Opin Genet Dev 2006;16: 578–585.
18 Orr HA, Kim Y: An adaptive hypothesis for the evolution of the Y chromosome. Genetics 1998; 150:1693–1698.
19 Charlesworth B, Sniegowski P, Stephan W: The evolutionary dynamics of repetitive DNA in eukaryotes. Nature 1994;371:215–220.
20 Pereira V: Insertion bias and purifying selection of retrotransposons in the *Arabidopsis thaliana* genome. Genome Biol 2004;5:R79.
21 Meyers BC, Tingey SV, Morgante M: Abundance, distribution, and transcriptional activity of repetitive elements in the maize genome. Genome Res 2001;11:1660–1676.
22 Sakamoto K, Abe T, Matsuyama T, Yoshida S, Ohmido N, et al: RAPID markers encoding retrotransposable elements are linked to the male sex in *Cannabis sativa* L. Genome 2005;48:931–936.
23 Kejnovsky E, Kubat Z, Macas J, Hobza R, Mracek J, Vyskot B: Retand: a novel family of gypsy-like retrotransposons harboring an amplified tandem repeat. Mol Genet Genomics 2006;276: 254–263.
24 Mariotti B, Navajas-Pérez R, Lozano R, Parker JS, de la Herrán R, et al: Cloning and characterization of dispersed repetitive DNA derived from microdissected sex chromosomes of *Rumex acetosa*. Genome 2006;49:114–121.

25 Obara M, Matsunaga S, Nakao S, Kawano S: A plant Y chromosome-STS marker encoding a degenerate retrotransposon. Genes Genet Syst 2002;77:393–398.
26 Liu Z, Moore PH, Ma H, Ackerman CM, Ragiba M, et al: A primitive Y chromosome in papaya marks the beginning of sex chromosome evolution. Nature 2004;427:348–352.
27 Sanmiguel P, Tikhonov A, Jin YK, Motchoulskaia N, Zakharov D, et al: Nested retrotransposons in the intergenic regions of the maize genome. Science 1997;274:765–768.
28 Naito K, Cho E, Yang GN, Campbell MA, Yano K, et al: Dramatic amplification of a rice transposable element during recent domestication. Proc Natl Acad Sci USA 2006;103: 17620–17625.
29 Ye D, Installé P, Ciuperescu C, Veuskens J, Wu Y, et al: Sex determination in the dioecious *Melandrium.* I. First lessons from androgenic haploids. Sex Plant Repr 1990;3:179–186.
30 Lardon A, Georgiev S, Aghmir A, Merrer GL, Negrutiu I: Sexual dimorphism in white campion: complex control of carpel number is revealed by Y chromosome deletions. Genetics 1999;151: 1173–1185.
31 Janousek B, Grant SR, Vyskot B: Non-transmissibility of the Y chromosome through the female line in androhermaphrodite plants of *Melandrium album*. Heredity 1998;80:576–583.
32 Guttman DS, Charlesworth D: An X-linked gene has a degenerate Y-linked homologue in the dioecious plant *Silene latifolia*. Nature 1998;393:263–266.
33 Jamsari A, Nitz I, Reamon-Buttner SM, Jung C: BAC-derived diagnostic markers for sex determination in asparagus. Theor Appl Genet 2004;108:1140–1146.
34 Kejnovsky E, Kubat Z, Hobza R, Lengerova M, Sato S, et al: Accumulation of chloroplast DNA sequences on the Y chromosome of *Silene latifolia*. Genetica 2006;128:167–175.
35 Delichère C, Veuskens J, Hernould M, Barbacar N, Mouras A, et al: SlY1, the first active gene cloned from a plant Y chromosome, encodes a WD-repeat protein. EMBO J 1999;18:4169–4179.
36 Atanassov I, Delichère C, Filatov DA, Charlesworth D, Negrutiu I, Monéger F: A putative monofunctional fructose-2,6-bisphosphatase gene has functional copies located on the X and Y sex chromosomes in white campion (*Silene latifolia*). Mol Biol Evol 2001;18:2162–2168.
37 Filatov DA: Substitution rates in a new *Silene latifolia* sex-linked gene, SlssX/Y. Mol Biol Evol 2005;22:402–408.
38 Nicolas M, Marais G, Hykelova V, Janousek B, Laporte V, et al: A gradual process of recombination restriction in the evolutionary history of the sex chromosomes in dioecious plants. PLoS Biol 2005;3:47–56.
39 Bergero R, Forrest A, Kamau E, Charlesworth D: Evolutionary strata on the X chromosomes of the dioecious plant *Silence latifolio*: evidence from new sex-linked genes. Genetics 2007;175: 1945–1954.
40 Lahn BT, Page DC: Four evolutionary strata on the human X chromosome. Science 1999;286: 964–967.
41 Skaletsky H, Kuroda-Kawaguchi T, Minx PJ, Cordum HS, Hillier L, et al: The male-specific region of the human Y chromosome is a mosaic of discrete sequence classes. Nature 2003;423: 825–837.
42 Waters PD, Duffy B, Frost CJ, Delbridge ML, Graves JAM: The human Y chromosome derives largely from a single autosomal region added to the sex chromosomes 80–130 million years ago. Cytogenet Cell Genet 2001;92:74–79.
43 Filatov DA, Charlesworth D: Substitution rates in the X- and Y-linked genes of the plants, *Silene latifolia* and *S. dioica*. Mol Biol Evol 2002;19:898–907.
44 Honys D, Twell D: Comparative analysis of the *Arabidopsis* pollen transcriptome. Plant Physiol 2003;132:640–652.
45 Nei M: Accumulation of nonfunctional genes on sheltered chromosomes. Am Nat 1970;104: 311–322.
46 Yu Q, Hou S, Feltus A, Jones MR, Murray JE, et al: Low X/Y divergence in four pairs of papaya sex-linked genes. Plant J 2008;53:124–132.
47 Yamato KT, Ishizaki K, Fujisawa M, Okada S, Nakayama S, et al: Gene organization of the liverwort Y chromosome reveals distinct sex chromosome evolution in a haploid system. Proc Natl Acad Sci USA 2007;104: 6472–6477.
48 Desfeux C, Maurice S, Henry JP, Lejeune B, Gouyon PH: Evolution of reproductive systems in the genus *Silene*. Proc R Soc Lond B 1996;263:409–414.

49 Zluvova J, Janousek B, Negrutiu I, Vyskot B: Comparison of the X and Y chromosome organization in *Silene latifolia*. Genetics 2005;170:1431–1434.
50 Navajas-Perez R, Schwarzacher T, de la Herrán R, Ruiz Rejón C, Ruiz Rejón M, Garrido-Ramos MA: The origin and evolution of the variability in a Y-specific satellite-DNA of *Rumex acetosa* and its relatives. Gene 2006;368:61–71.
51 Mosiolek M, Pasierbek P, Malarz J, Mos M, Joachimiak A: *Rumex acetosa* Y chromosomes: constitutive or facultative heterochromatin? Folia Histochem Cytobiol 2005;43:161–167.
52 Matsunaga S, Isono E, Kejnovsky E, Vyskot B, Kawano S, Charlesworth D: Duplicative transfer of a MADS box gene to a plant Y chromosome. Mol Biol Evol 2003;20:1062–1069.
53 Marin J, Baker BS: The evolutionary dynamics of sex determination. Science 1998;281: 1990–1994.
54 Spada A, Caporali E, Marziani G, Portaluppi P, Restivo FM, et al: A genetic map of *Asparagus officinalis* based on integrated RFLP, RAPD and AFLP molecular markers. Theo Appl Genet 1998; 97:1083–1089.
55 Harvey CF, Gill GP, Fraser LG, McNeilage MA: Sex determination in *Actinidia*. 1. Sex-linked markers and progeny sex ratio in diploid *A. chinensis*. Sex Plant Reprod 1997;10:149–154.
56 Chiang Y-C, Schaal BA, Chou C-H, Huang S, Chiang T-Y: Contrasting selection modes at the Adh1 locus in outcrossing *Miscanthus sinensis* vs. inbreeding *Miscanthus condensatus* (Poaceae). Am J Bot 2003;90:561–570.
57 Dalbó MA, Ye GN, Weeden NF, Steinkellner H, Sefc KM, Reisch BI: A gene controlling sex in grapevines placed on a molecular marker-based genetic map. Genome 2000;43:333–340.
58 Khattak J, Torp A, Andersen S: A genetic linkage map of *Spinacia oleracea* and localization of a sex determination locus. Euphytica 2006;148:311–318.
59 Peil A, Flachowsky H, Schumann E, Weber WE: Sex-linked AFLP markers indicate a pseudoautosomal region in hemp (*Cannabis sativa* L.). Theo Appl Genet 2003;107:102–109.
60 Rode J, In-Chol K, Saal B, Flachowsky H, Kriese U, Weber WE: Sex-linked SSR markers in hemp. Plant Breeding 2005;124:167–170.
61 Scotti I, Delph LF: Selective trade-offs and sex-chromosome evolution in *Silene latifolia*. Evolution 2006;60:1793–1800.
62 Seefelder S, Ehrmaier H, Schweizer G, Seigner E: Male and female genetic linkage map of hops, *Humulus lupulus*. Plant Breeding 1999;119:249–255.

Deborah Charlesworth
Institute of Evolutionary Biology, School of Biological Sciences
University of Edinburgh, Ashworth Lab. King's Buildings
W. Mains Rd., Edinburgh EH9 3JT (UK)
Tel. +44 131 650 5751, Fax +44 131 650 6564, E-Mail deborah.charlesworth@ed.ac.uk

Volff J-N (ed): Plant Genomes.
Genome Dyn. Basel, Karger, 2008, vol 4, pp 95–107

Plant Centromeres

J.C. Lamb, W. Yu, F. Han, J.A. Birchler

Division of Biological Sciences, University of Missouri-Columbia,
Columbia, Mo., USA

Abstract

Plant centromeres are generally composed of tandem arrays of simple repeats that are typical of a particular species, but that evolve rapidly. Centromere specific retroelements are also present. These arrays associate with a centromere specific variant of histone H3 that anchors the site of the kinetochore. Although such DNA arrays are typical of the centromere, the specification of centromere activity has an epigenetic component as shown by the fact that centromeres are formed in the absence of such repeats and that centromeres in dicentric chromosomes regularly undergo inactivation.

The centromere is the portion of the chromosome that partitions the sister chromatids during mitosis. In meiosis, it holds together the sisters at meiosis I and then partitions them during meiosis II. Long thought to be straightforward in their function, recent data describing the evolution of centromeric DNA and the specification of the chromatin structure associated with that DNA show that many of the basic properties of centromeres remain unexplained. In this review, we discuss recent developments in research on plant centromeres.

Typically, plant centromeres are composed of short repetitive sequences arranged in complex arrays often together with a centromere specific retrotransposon [1]. These sequences evolve quite rapidly and have little similarity across distant taxonomic relationships in the kingdom although the kinetochore proteins are highly conserved. The size of the repetitive blocks can be significantly reduced while maintaining centromere activity until a lower threshold is achieved [2–4]. The site of the kinetochore is anchored on the chromosome by the deposition of a variant of histone H3 called CenH3. Although CenH3 deposition is typically coincident with a portion of the centromeric repeats, the DNA sequence alone is neither necessary nor sufficient to determine the location of the kinetochore.

Chromatin Structure at the Centromere

The presence of a variant of the H3 histone protein, CenH3, found only at centromeres is the defining feature of chromatin that composes the kinetochore forming domain. At the centromere, the CenH3 protein replaces the canonical H3 histone protein forming the foundation of the multilevel kinetochore complex. Because the CenH3 containing nucleosomes are associated with the DNA, they likely provide a physical link between microtubules and chromosomes. The centromere region also keeps the sister chromatids attached until all chromosomes are aligned at the metaphase plate. The sister kinetochores are physically separated before microtubule attachment [5], and in yeast and mammals this separation has been shown to depend on a number of other factors [5]. The centromere is separated into at least two distinct parts during metaphase, one enriched for CenH3 and the other associated with cohesins, which hold the sister chromatids together. Characterizing the chromatin structure of these centromere regions could provide clues to how they are formed and function.

The DNA at and around centromeres is typically composed of long tracts of satellite repeats arrayed head to tail. The length of these tracts can be considerable, thousands of repeats and millions of base pairs. Since repeat arrays are a common feature of centromeres, it is likely that they represent an optimal DNA substrate to form the centromere structure. However, since the arrays are dispensable for centromere formation, the epigenetic features that determine centromeres must not strictly depend on them. Therefore, care must be taken when interpreting observations of chromatin structure of centromeric regions to distinguish between features that are solely a result of repeat arrays and those features that represent true hallmarks of centromere identity. One way to separate these two properties is to examine centromeres present on chromosomal regions with few or no repeats.

Yan and coworkers [6] took advantage of the small size of the centromeric repeat tract and complete sequence of rice centromere 8 to examine the chromatin state at a centromere without a large repeat array. They used chromatin immuno-precipitation with antibodies against various histone modifications followed by PCR (ChIP-PCR) with primers specific to the centromere regions to determine the presence of histone modifications at many positions within the region of recombination repression [6]. This region contained relatively few centromeric repeats and, near the edges of the region, a similar ratio of retrotransposons as the rest of the genome. It also contained many expressed genes. Here, as in other genomic regions, active genes were marked with H3K4me2 and H4-acetylation while the remainder of the chromatin was mostly associated with H3K9me2. The authors conclude that gross histone modifications do not determine centromere identity and imply that previous observations of distinct patterns of histone modifications at centromeres [7, 8] merely reflect the unusual density of DNA repeats found there.

The authors further conclude that repression of recombination is not due to the gene poor, or repeat rich nature of the pericentromeric region but is an intrinsic property of centromeres [6]. They suggest that cohesin proteins are recruited by interactions with CenH3 and that their abundance in the pericentromeric region prevents recombination. Thus, repression of recombination is an epigenetic consequence of centromere activity. This model takes into account observations that relocated centromeres repress recombination in the juxtaposed regions [9].

In plants, the region between the separating kinetochores is marked by phosphorylation of the H3 histone at the Serine 10 and Serine 28 residues, and this mark is thought to indicate the regions involved in sister chromatid cohesion [10]. Although H3 phosphorylation does not itself cause cohesion [11], it probably does reflect an epigenetic marking of the chromatin that accumulates and retains high levels of cohesins. In maize, phosphorylation first appears on the CenH3 protein and then spreads to H3 proteins in the flanking region, suggesting that phosphorylation is directed by CenH3 position to the surrounding regions [12]. In *Luzula luzuloides*, centromere activity and the CenH3 protein [13] as well as phosphorylation of H3 [11] is distributed along the whole chromosome rather than at a single location. Even in this somewhat unusual case, the pattern is consistent with CenH3 directed H3 phosphorylation.

In rice [14] as well as humans and flies [7, 8], the CenH3 nucleosomes are interspersed with nucleosomes containing canonical H3 histones. In all plant species where anti-CenH3 antibodies are available, immuno-labeling shows that the centromere repeat arrays are larger than the CenH3 containing chromatin domain [13–17]. In *Arabidopsis* and maize, the excess centromere repeat is associated with phosphorylated H3 and many repeats are located between the sister kinetochores at the site of chromosome cohesion in metaphase [10, 12]. Therefore, the cohesion and kinetochore domains may be interspersed and the chromatin folded or looped so that the repeats that are wrapped around the canonical nucleosomes are on the inside of the centromere and the CenH3 associated repeats face outward. Such an arrangement would place the two domains in intimate contact until separation at G2, thus facilitating CenH3 directed assembly of the chromatin state required for chromatid cohesion.

Attributing recruitment of cohesins, and indirectly the formation of the recombination free zone, to CenH3, provides a model that explains how the locations of these centromere properties are determined. This model also highlights the fundamental question of centromere biology: How is the position of the CenH3 binding site determined and maintained?

Epigenetic Determination of Centromere Identity

Recently, direct evidence demonstrating the insufficiency of repeats for plant centromere determination has been uncovered that closely parallel observations made in other model systems. First, an example of a new centromere devoid of any centromeric elements was presented [18]. Chromosomal mutations were induced on a barley chromosome in a wheat background and among the variations recovered were two isochromosomes that had a functional centromere but no centromeric repeats. Because the centromere of the isochromosome lacked the centromeric repeats found on its progenitor, it must have formed de novo at a location previously without centromere activity. If such de novo centromere formation events have occurred recently, the sequence at the new centromere would initially appear similar to other, non-centromeric, regions and would gradually acquire centromere repeats. Two rice centromeres that have little centromere specific DNA have been completely sequenced and show such a pattern. Numerous genes are present in and around the centromere regions, consistent with a relatively recent centromere formation event [6].

Indeed, an epigenetic component to centromere specification was established for plant centromeres by the observation of efficient inactivation of a maize centromere [19]. Using a dispensable chromosome containing an inverted duplication, a dicentric chromosome was produced that entered the chromosome type breakage-fusion-bridge (BFB) cycle. The dicentric chromosome was ripped apart during subsequent mitotic division producing minichromosomes. The BFB cycle may be ended if one of the centromeres is inactivated so the collection of minichromosomes was examined for dicentrics. In addition to dicentric minichromosomes, the inactivated centromere was also found affixed to the end of chromosome 9 where it remained inactive [19]. Because the primary DNA sequence does not determine centromere identity, epigenetic factors, such as chromatin structure, must be a major determinant. In this regard, plant centromeres are similar to other eukaryotic centromeres. Recent identification of additional proteins in plants, including aurora kinases [20] and MIS12 [21], which are homologous to essential mammalian and yeast proteins, further emphasizes the high degree of conservation for structure and function of eukaryotic centromeres.

If CenH3 is not directed to the kinetochore by the underlying sequence, how is its location determined? Some clues have been provided by a recent study of CenH3 incorporation in *Arabidopsis* [5]. Examining intensity of immunolabeled *Arabidopsis* CenH3 at different points in the cell cycle suggests that CenH3 loading occurs primarily during G2 [5]. CenH3 intensity increased at the same time that the CenH3 domains of the sister centromeres were separated, suggesting that CenH3 incorporation and kinetochore formation occur at the same time. CenH3 quantity was also examined in endoreduplicated cells, where DNA replication has occurred without subsequent cell division. The amount of CenH3 did not increase proportionally to the nuclear DNA. Assuming that centromeric DNA was not under-replicated compared to the rest of the genome, this result shows that replication coupled CenH3 deposition does not occur or is insufficient to maintain CenH3 levels in these cells [5].

Two possibly complementary models for CenH3 maintenance include: 1) new CenH3 incorporation is targeted to the kinetochore forming region by the CenH3 that is already there and 2) CenH3 incorporation can occur indiscriminately but is excluded from most chromosomal locations except the centromere. The CenH3 loading complex appears to be very simple, unlike that of nucleosomes containing the canonical H3 histone, and is sufficient to incorporate CenH3 in vitro [22, 23]. The simplicity of the complex is consistent with replication independent incorporation. Another possibility is that incorporation may be directed by interactions with CenH3 already present on the chromosome or other centromere proteins. The increase in CenH3 levels in *Arabidopsis* may occur simultaneously with kinetochore formation [5]. If so, then kinetochore proteins bound by CenH3 that is already on the chromosome may in turn recruit free CenH3-nucleosomes leading to maintenance of CenH3 levels. Mis-incorporated CenH3 may be eliminated by proteolysis. In *S. pombe* and *Drosophila* CenH3 levels at the centromere were shown to be regulated by protein degradation and failure of CenH3 proteolysis led to ectopic incorporation [24, 25].

Role of Centromere Repeats in Determining Centromere Identity

Although centromeric repeats are neither sufficient nor necessary for centromere function, they are present exclusively at and around the kinetochore. This suggests a role, or at least a preference, for repeats in centromere function [1]. Alternatively, centromere repeats are a by-product of centromere function. One likely possibility is that the repeat arrays are optimal for formation of the correct chromatin conformation at the centromere and complement the epigenetic determinants of centromere identity [26]. Repeat optimization could be for either, or both, of the two major functions of centromeres, movement and cohesion.

Many recent studies have identified centromere repeats in different plant species allowing repeats among related taxa to be compared and providing new insights into the mechanisms of centromere repeat evolution. Assuming changes in centromere repeats are driven by their optimization for centromere function, the nature of repeat evolution can provide insight into how centromere identity is determined.

Many, but not all, of the relatives of rice, the genus *Oryza*, share a related centromeric satellite, CentO [27]. Using chromatin immunoprecipitation, three repeats were identified at the kinetochores of *O. rhizomatis* and *O. brachyantha*, two relatives without CentO. Two of these repeats had an 80-bp region with homology to CentO as well as to the centromere repeats of maize and millet [27]. The third repeat, composed mostly from the *O. brachyantha* centromere was totally novel. Thus, conservation of some sequences from rice to maize was detected but some lineages have acquired new repeats. In *O. officinalis* and *O. brachyantha*, CentO and another centromere repeat, CentO-C, are both present at some centromeres and one or the other element is present in the remainder [28]. Also, in *O. brachyantha*, several centromeres contained a 366-bp repeat, TrsC, that is also present in the subtelomeric region. TrsC is only found in the subtelomeric region of *O. officinalis* [28]. These patterns of centromeric elements demonstrate that repeats can rapidly change, either by alterations to existing elements or by development of new elements including by recruitment from other genomic regions. The species with multiple repeats may represent a transition from one predominant type to another.

Among *Arabidopsis* relatives, variation is also observed for centromeric elements [29, 30]. In *A. gemmifera*, several centromere repeats are present in different combinations among the various centromeres. Additionally, variation for repeat composition is seen for homologous centromeres among different individuals of this species. The authors concluded that the repeats have been rearranged as blocks among the different centromeres [30]. In *A. halleri* and *A. lyrata*, species with multiple centromeric element families, two CenH3 homologs are present, and one is polymorphic among the populations examined [31]. Although the multiple repeats and multiple proteins may be a related phenomenon, additional species will need to be identified with duplicate CenH3 (or other centromere proteins) genes and their centromere sequence composition determined before the importance of this finding can be evaluated.

The length of the centromere satellite in many eukaryotes is approximately the length required to wrap around one or two nucleosomes [32]. The potential one-to-one association of centromere repeats and centromeric nucleosomes may be conducive to forming a higher-order chromatin structure. The majority, but not all, of plant centromeric satellites that have been cloned [e.g. 13, 27, 28] fit this pattern. In contrast, large blocks of a tri-nucleotide microsatellite, $(CAA)_n$, were recently identified at the centromeres of most, but not all tomato centromeres [33]. This arrangement is similar to *Drosophila* centromeres, which contain two different 5-mer microsatellites [34]. Seven base pair telomeric satellites have also been identified at or near centromeres in tomato [35] and in the closely related species, potato [36]. These smaller repeats may reflect a different type of association between DNA and nucleosomes. Alternatively, a set number of the smaller repeats may form a larger unit to form the specialized chromatin structure.

Careful analysis of the centromere 8 region of rice showed a high frequency of solo LTRs in the CenH3 binding domain [37]. One mechanism of solo-LTR formation is unequal recombination. Given that meiotic recombination is repressed around centromeres, the mechanism of such recombination is unclear. Further analysis, including comparisons of this region between *indica* and *japonica* rice identified large duplicated blocks and rearranged segments of centromere repeats [37]. Centromeric retroelements found at the centromere accounted for a large portion of the total sequence but element number was increased primarily by segmental duplications instead of transposition [38]. Thus, both within and among centromeres, duplications or exchanges of large blocks of sequence are responsible for the observed variation.

Given that centromere determination has an epigenetic component, the optimization of repeats for centromere chromatin formation may serve to reinforce a centromere location. Much attention has focused on the association of centromeric repeats with CenH3 at the kinetochore, with the implicit assumption that repeats are optimized for kinetochore functions. Because centromere satellites in many species are approximately the size expected to associate with one nucleosome, optimization likely involves the repeat length and may serve to create arrays of regularly positioned centromeric nucleosomes [32]. Both the CenH3 protein and another essential centromere protein, CenpC, are hypervariable for the part of the proteins that contacts DNA. This variation in centromere proteins could be related to the variation for centromeric DNA repeats. Also, within centromere arrays, the elements at the kinetochore appear to be the newest because they are the most homogeneous. The association of the newest variants with the functional kinetochore suggests that variation is produced at the kinetochore or that the site of kinetochore formation can move to associate with and stabilize the most favorable repeats. There has been less emphasis on possible roles of these repeats in chromatid cohesion. Recruitment or stabilization of cohesins could conceivably be influenced by the binding or folding properties of the underlying DNA sequence or by the chromatin state.

A common feature of the centromeric region is the presence of heterochromatin flanking the kinetochore. In plants, this pericentromeric heterochromatin can extend well into the arms away from the functional centromere. Heterochromatin may have properties that are valuable, although dispensable, for centromere function and chromosome distribution. In particular, it has been suggested that heterochromatin plays a role in chromosome cohesion. In fission yeast, the repeats flanking the kinetochore are essential for sister chromatid cohesion and are maintained in a proper heterochromatic state by the RNAi machinery [39, 40]. Similarly, pericentromeric heterochromatin is required for proper chromosome cohesion and disjunction in flies and other systems [41]. In chicken cells, RNAi components are required for proper centromere function [42].

However, in plants, mutations that affect heterochromatin formation do not lead to chromosome missegregation [reviewed in 43]. In meiosis, recombination near the centromere could increase the likelihood that sister centromeres would separate leading to missegregation. Although heterochromatin is present in the region of repressed recombination, the heterochromatin state itself does not appear to be the cause [6]. Because recombination is infrequent in the pericentromeric regions, repetitive elements including retrotransposons accumulate leading to heterochromatin formation [43]. Thus, instead of being a cause, the pericentromeric heterochromatin might be a byproduct of repressed recombination near the centromere.

Although not essential for centromere function, the creation of heterochromatic regions around centromeres could optimize chromosomes for transmission and provide genome stability. Proximal heterochromatin could serve as a spacer that distances the genetic component of a chromosome from the functional component [1]. For example, moving genes away from the centromere would ensure their participation in the allelic shuffling of meiosis. Because cohesin is more abundant in heterochromatin than euchromatin, perhaps through association with HP1 [44], the cohesin molecules may be more readily available for organization by CenH3 into a chromatid cohesion complex. Heterochromatin could also provide boundaries for CenH3 deposition, perhaps by maintaining a closed conformation that is inaccessible to CenH3 loading machinery. Finally, after incorporation, heterochromatin may protect CenH3 molecules from proteolysis.

It is worth noting that the heterochromatin around centromeres consists of various repetitive elements and even genes, but immediately adjacent to the kinetochore, there are tandem arrays of centromeric repeats that are involved in chromatid cohesion. Tandem arrays of repeats have intrinsic heterochromatin forming abilities. Regular spacing of nucleosomes allows for highly regular packaging. Low levels of expression of repeat arrays have been postulated to generate a self perpetuating population of small RNA molecules that could maintain a heterochromatin state [45]. Indeed, centromeric elements have been shown to be expressed at low levels in maize [46], rice [16] and *Arabidopsis* [47]. Thus, while additional alterations to repeat units or arrangements may allow further optimization for association with and organization by centromeric elements, a large region of repeats may generally be conducive to centromere formation.

Neocentromeres and De Novo Centromeres

Neocentromeres are a class of activated centromeres that have formed at sites on chromosomes that typically do not have centromeric activity and that lack the canonical centromeric arrays. Neocentromeres can form in mammals and flies as functional units to substitute for the original centromeres, whose functions were lost due to mutation or deletions [48, 49]. Unlike the neocentromeres in human and flies which contain the same features as typical centromeres, a class of neocentromeres was originally defined in plants that only operate in meiosis when they become attached to spindle fibers and move ahead of the typical centromeres to the poles. The pulling of the chromosome with a neocentromere causes a meiotic drive, which results in the favorable transmission of these chromosomes [50, 51]. In maize, the neocentromere activity is determined by two tandem repeat arrays, the 180-bp heterochromatin knob and the 350-bp TR-1, and is controlled by abnormal chromosome 10 (Ab10) [52, 53]. Unlike the counterparts in human and flies, the plant neocentromeres are not associated with the centromere proteins such as CenH3 [50].

De novo functional centromeres are formed when DNA including centromeric and other necessary sequences is introduced into cells to form a functional chromosome. For example, the *Saccharomyces cerevisiae* centromere sequence can be as small as 125 bp, and by using a 107-bp CEN4 sequence from chromosome 4, functional yeast artificial chromosome vectors can be built and used for the cloning of large pieces of genomic DNA [54, 55]. As compared with the simple sequence requirement of yeast, de novo human centromere formation requires millions of basepairs of centromeric array [56]. The centromeric repeats of many plant species have been cloned including those described above and others. However, re-introduction of plant centromeric repeat arrays has not produced de novo centromeres.

Recently, two BAC clones of centromere-specific DNAs were transformed into rice by bombardment in an effort to produce de novo plant artificial chromosomes [57]. Although transformants were successfully regenerated, all the transgenics with the centromere BACs were integration events. A comparison of transformation with centromere BACs and genomic BACs revealed a lower frequency of transformation and regeneration of transgenic calli using the centromere BACs. Additionally, some phenotypic abnormalities in transgenic plants containing the centromere BACs were reported. One explanation is that de novo centromere formation occurred at an early stage of the transcentromere events, but the centromere was subsequently inactivated by a similar mechanism to that observed in dicentric maize chromosomes [19]. There are several possible explanations for why de novo centromere formation failed in rice but was successful in human cells. First, the centromeric DNA size in the transformed plants was much smaller than at a normal centromere; a certain threshold might be necessary. The size of the centromere repeat array may have been further reduced by fragmentation or rearrangement resulting from the plant transformation process. In mammalian cells, it has been found that the size of centromeric DNA has a positive relationship with the efficiency of de novo artificial chromosome formation [58]. Second, other sequences may be necessary. For example, transfection with human alphoid sequence without the CENP-B binding site significantly reduced the efficiency of human artificial chromosome formation [59]. Third, in human de novo centromere construction, a cultured cell line was used, while in plants a putative de novo centromere will have to endure the processes of dedifferentiation and regeneration during which it may be selected against.

The practical aim of developing de novo centromeres is the production of plant artificial chromosomes. Such chromosomes could be used as a platform to receive multiple transgenes allowing the introduction of multigenic traits. An alternative to de novo centromere formation is to create an artificial chromosome by manipulating a natural chromosome. By introducing transgenic telomere repeat arrays, chromosomes can be truncated [60]. Therefore, telomere mediated chromosomal truncation provides a solution to the current inability to recover de novo centromeres in plants.

Acknowledgements

Work on this topic in our laboratory is funded by a grant from the National Science Foundation Plant Genome Initiative (DBI 0421671).

References

1 Lamb JC, Theuri J, Birchler JA: What's in a centromere? Genome Biol 2004;5:239.
2 Kaszas E, Birchler J: Misdivision analysis of centromere structure in maize. EMBO J 1996;15:5246–5255.
3 Kaszas E, Birchler J: Meiotic transmission rates correlate with physical features of rearranged centromeres in maize. Genetics 1998;150:1683–1692.
4 Jin W, Lamb JC, Vega JM, Dawe RK, Birchler JA, Jiang J: Molecular and functional dissection of the maize B chromosome centromere. Plant Cell 2005;17:1412–1423.
5 Lermontova I, Schubert V, Fuchs J, Klatte S, Macas J, Schubert I: Loading of *Arabidopsis* centromeric histone CENH3 occurs mainly during G2 and requires the presence of the histone fold domain. Plant Cell 2006;18:2443–2451.
6 Yan H, Jin W, Nagaki K, Tian S, Ouyang S, et al: Transcription and histone modifications in the recombination-free region spanning a rice centromere. Plant Cell 2005;17:3227–3238.
7 Blower MD, Sullivan BA, Karpen GH: Conserved organization of centromeric chromatin in flies and humans. Dev Cell 2002;2:319–330.
8 Sullivan BA, Karpen GH: Centromeric chromatin exhibits a histone modification pattern that is distinct from both euchromatin and heterochromatin. Nat Struct Mol Biol 2004;11:1076–1083.
9 Choo KH: Why is the centromere so cold? Genome Res 1998;8:81–82.
10 Shibata F, Murata M: Differential localization of the centromere-specific proteins in the major centromeric satellite of *Arabidopsis thaliana*. J Cell Sci 2004;117:2963–2970.

11 Gernand D, Demidov D, Houben A: The temporal and spatial pattern of histone H3 phosphorylation at serine 28 and serine 10 is similar in plants but differs between mono- and polycentric chromosomes. Cytogenet Genome Res 2003;101:172–176.
12 Zhang X, Li X, Marshall JB, Zhong CX, Dawe RK: Phosphoserines on maize CENTROMERIC HISTONE H3 and histone H3 demarcate the centromere and pericentromere during chromosome segregation. Plant Cell 2005;17:572–583.
13 Nagaki K, Murata M: Characterization of CENH3 and centromere-associated DNA sequences in sugarcane. Chromosome Res 2005;13:195–203.
14 Nagaki K, Cheng Z, Ouyang S, Talbert PB, Kim M, et al: Sequencing of a rice centromere uncovers active genes. Nat Genet 2004;36:138–145.
15 Nagaki K, Song J, Stupar RM, Parokonny AS, Yuan Q, et al: Molecular and cytological analyses of large tracks of centromeric DNA reveal the structure and evolutionary dynamics of maize centromeres. Genetics 2003;163:759–770.
16 Zhang W, Yi C, Bao W, Liu B, Cui J, et al: The transcribed 165-bp CentO satellite is the major functional centromeric element in the wild rice species *Oryza punctata*. Plant Physiol 2005;139: 306–315.
17 Zhong CX, Marshall JB, Topp C, Mroczek R, Kato A, et al: Centromeric retroelements and satellites interact with maize kinetochore protein CENH3. Plant Cell 2002;14:2825–2836.
18 Nasuda S, Hudakova S, Schubert I, Houben A, Endo TR: Stable barley chromosomes without centromeric repeats. Proc Natl Acad Sci USA 2005;102:9842–9847.
19 Han F, Lamb JC, Birchler JA: High frequency of centromere inactivation resulting in stable dicentric chromosomes of maize. Proc Natl Acad Sci USA 2006;103:3238–3243.
20 Kawabe A, Matsunaga S, Nakagawa K, Kurihara D, Yoneda A, et al: Characterization of plant Aurora kinases during mitosis. Plant Mol Biol 2005;58:1–13.
21 Sato H, Shibata F, Murata M: Characterization of a Mis12 homologue in *Arabidopsis thaliana*. Chromosome Res 2005;13:827–834.
22 Furuyama T, Dalal Y, Henikoff S: Chaperone-mediated assembly of centromeric chromatin in vitro. Proc Natl Acad Sci USA 2006;103:6172–6177.
23 Furuyama T, Henikoff S: Biotin-tag affinity purification of a centromeric nucleosome assembly complex. Cell Cycle 2006;5:1269–1274.
24 Collins KA, Furuyama S, Biggins S: Proteolysis contributes to the exclusive centromere localization of the yeast Cse4/CENP-A histone H3 variant. Curr Biol 2004;14:1968–1972.
25 Moreno-Moreno O, Torras-Llort M, Azorin F: Proteolysis restricts localization of CID, the centromere-specific histone H3 variant of *Drosophila*, to centromeres. Nucleic Acids Res 2006;34: 6247–6255.
26 Dawe RK, Henikoff S: Centromeres put epigenetics in the driver's seat. Trends Biochem Sci 2006;31:662–669.
27 Lee HR, Zhang W, Langdon T, Jin W, Yan H, et al: Chromatin immunoprecipitation cloning reveals rapid evolutionary patterns of centromeric DNA in *Oryza* species. Proc Natl Acad Sci USA 2005; 102:11793–11798.
28 Bao W, Zhang W, Yang Q, Zhang Y, Han B, et al: Diversity of centromeric repeats in two closely related wild rice species, *Oryza officinalis* and *Oryza rhizomatis*. Mol Genet Genomics 2006;275:421–430.
29 Kawabe A, Nasuda S: Structure and genomic organization of centromeric repeats in *Arabidopsis* species. Mol Genet Genomics 2005;272:593–602.
30 Kawabe A, Nasuda S: Polymorphic chromosomal specificity of centromere satellite families in *Arabidopsis halleri* ssp. *gemmifera*. Genetica 2006;126:335–342.
31 Kawabe A, Nasuda S, Charlesworth D: Duplication of centromeric histone H3 (HTR12) gene in *Arabidopsis halleri* and *lyrata*, plant species with multiple centromeric satellite sequences. Genetics 2006;174:2021–2032.
32 Henikoff S, Ahmad K, Malik HS: The centromere paradox: stable inheritance with rapidly evolving DNA. Science 2001;293:1098–1102.
33 Yang TJ, Lee S, Chang SB, Yu Y, de Jong H, Wing RA: In-depth sequence analysis of the tomato chromosome 12 centromeric region: identification of a large CAA block and characterization of pericentromere retrotranposons. Chromosoma 2005;114:103–117.

34 Sun X, Le HD, Wahlstrom JM, Karpen GH: Sequence analysis of a functional *Drosophila* centromere. Genome Res 2003;13:182–194.
35 Presting GG, Frary A, Pillen K, Tanksley SD: Telomere-homologous sequences occur near the centromeres of many tomato chromosomes. Mol Gen Genet 1996;251:526–531.
36 Tek AL, Jiang J: The centromeric regions of potato chromosomes contain megabase-sized tandem arrays of telomere-similar sequence. Chromosoma 2004;113:77–83.
37 Ma J, Bennetzen JL: Recombination, rearrangement, reshuffling, and divergence in a centromeric region of rice. Proc Natl Acad Sci USA 2006;103:383–388.
38 Ma J, Jackson SA: Retrotransposon accumulation and satellite amplification mediated by segmental duplication facilitate centromere expansion in rice. Genome Res 2006;16:251–259.
39 Bernard P, Maure JF, Partridge JF, Genier S, Javerzat JP, Allshire RC: Requirement of heterochromatin for cohesion at centromeres. Science 2001;294:2539–2542.
40 Volpe T, Schramke V, Hamilton GL, White SA, Teng G, et al: RNA interference is required for normal centromere function in fission yeast. Chromosome Res 2003;11:137–146.
41 Bernard P, Allshire R: Centromeres become unstuck without heterochromatin. Trends Cell Biol 2002;12:419–424.
42 Fukagawa T, Nogami M, Yoshikawa M, Ikeno M, Okazaki T, et al: Dicer is essential for formation of the heterochromatin structure in vertebrate cells. Nat Cell Biol 2004;6:784–791.
43 Topp CN, Dawe RK: Reinterpreting pericentromeric heterochromatin. Curr Opin Plant Biol 2006;6:647–653.
44 Nonaka N, Kitajima T, Yokobayashi S, Xiao G, Yamamoto M, et al: Recruitment of cohesin to heterochromatic regions by Swi6/HP1 in fission yeast. Nat Cell Biol 2002;4:89–93.
45 Martienssen RA: Maintenance of heterochromatin by RNA interference of tandem repeats. Nat Genet 2003;35:213–214.
46 Topp CN, Zhong CX, Dawe RK: Centromere-encoded RNAs are integral components of the maize kinetochore. Proc Natl Acad Sci USA 2004;101:15986–15991.
47 May BP, Lippman ZB, Fang Y, Spector DL, Martienssen RA: Differential regulation of strand-specific transcripts from *Arabidopsis* centromeric satellite repeats. PLoS Genet 2005;1:e79.
48 Amor DJ, Choo KH: Neocentromeres: role in human disease, evolution, and centromere study. Am J Hum Genet 2002;71:695–714.
49 Maggert KA, Karpen GH: The activation of a neocentromere in *Drosophila* requires proximity to an endogenous centromere. Genetics 2001;158:1615–1628.
50 Dawe RK, Hiatt EN: Plant neocentromeres: fast, focused, and driven. Chromosome Res 2004;12: 655–669.
51 Mroczek RJ, Melo JR, Luce AC, Hiatt EN, Dawe RK: The maize Ab10 meiotic drive system maps to supernumerary sequences in a large complex haplotype. Genetics 2006;174:145–154.
52 Hiatt EN, Kentner EK, Dawe RK: Independently regulated neocentromere activity of two classes of tandem repeat arrays. Plant Cell 2002;14:407–420.
53 Rhoades MM, Vilkomerson H: On the anaphase movement of chromosomes. Proc Natl Acad Sci USA 1942;28:433–436.
54 Burke DT, Carle GF, Olson MV: Cloning of large segments of exogenous DNA into yeast by means of artificial chromosome vectors. Science 1987;236:806–812.
55 Kuhn RM, Ludwig RA: Complete sequence of the yeast artificial chromosome cloning vector pYAC4. Gene 1994;141:125–127.
56 Harrington JJ, Van Bokkelen G, Mays RW, Gustashaw K, Willard HF: Formation of de novo centromeres and construction of first-generation human artificial microchromosomes. Nat Genet 1997;15:345–355.
57 Phan BH, Jin W, Topp CN, Zhong CX, Jiang J, et al: Transformation of rice with long DNA-segments consisting of random genomic DNA or centromere-specific DNA. Transgenic Res 2007;16:341–351.
58 Kaname T, McGuigan A, Georghiou A, Yurov Y, Osoegawa K, et al: Alphoid DNA from different chromosomes forms de novo minichromosomes with high efficiency. Chromosome Res 2005; 13:411–422.

59 Basu J, Stromberg G, Compitello G, Willard HF, Van Bokkelen G: Rapid creation of BAC-based human artificial chromosome vectors by transposition with synthetic alpha-satellite arrays. Nucleic Acids Res 2005;33:587–596.
60 Yu W, Lamb JC, Han F, Birchler JA: Telomere-mediated chromosomal truncation in maize. Proc Natl Acad Sci USA 2006;103:17331–17336.

Dr. James A. Birchler
Division of Biological Sciences
117 Tucker Hall, University of Missouri-Columbia
Columbia, MO 65211–7400 (USA)
Tel. +1 573 882 4905, Fax +1 573 882 0123, E-Mail birchlerj@missouri.edu

Volff J-N (ed): Plant Genomes.
Genome Dyn. Basel, Karger, 2008, vol 4, pp 108–118

miRNAs in the Plant Genome: All Things Great and Small

B.C. Meyers[a], *P.J. Green*[a,b], *C. Lu*[a]

[a]Department of Plant and Soil Sciences and Delaware Biotechnology Institute, and [b]College of Marine and Earth Sciences, University of Delaware, Newark, Del., USA

Abstract

Plants produce two major types of small RNAs that are 21 to 24 nucleotides in size. Small interfering RNAs (siRNAs) are typically involved in transcriptional gene silencing that results from the targeting of genomic DNA and triggering of histone modifications or DNA methylation. Deep sequencing experiments have demonstrated that thousands of loci, usually repetitive sequences, generate these siRNAs. In contrast, microRNAs (miRNAs) are encoded by perhaps just several hundred loci per genome that generate Pol II-derived single stranded precursors which are processed into specific miRNAs. miRNAs act in a post-transcriptional manner to regulate gene function. Recent work has focused on the identification and classification of small RNA-producing loci, as well as understanding small RNA targeting and function, and the evolution of this relatively recently discovered class of regulatory molecules.

Studies of microRNAs (miRNAs) and short-interfering RNAs (siRNAs) in plants have demonstrated that these molecules are potent and specific regulators of gene expression. Small RNAs have an important role in plant development, stress responses, and epigenetic regulation, primarily through their role in transcriptional and post-transcriptional silencing of specific target genes and other loci. While most studies of plant miRNAs focus on *Arabidopsis*, recent data on small RNA evolution suggest that miRNAs are components of a well-conserved gene regulatory system. miRNA-based regulation is a conserved feature of most eukaryotes, but high levels of sequence conservation for a small number of loci among flowering plants and mosses have demonstrated an ancient origin for individual miRNAs. It is likely that all plants utilize small RNA signaling networks for critical biological functions. Different plant families are likely to have both common and lineage-specific miRNAs or other small RNAs with important biological roles that have yet to be characterized.

Plant Small RNAs

Two major types of small RNAs (21 to 24 nucleotides in size), known as small interfering RNAs (siRNAs) [1], and microRNAs (miRNAs) [2], are present in a wide variety of eukaryotic organisms [3, 4]. Both types of molecules are processed from longer precursor RNAs by RNase III enzymes called DICERs [5–8]. Though siRNAs and miRNAs are similar in size and can both associate with Argonaute (AGO) proteins in effector complexes, their biogenesis and often their functions are substantially different [3, 9, 10]. siRNAs are processed from longer double stranded RNA (dsRNA) molecules and represent both strands of the encoding DNA molecules. In plants, transcriptional gene silencing can be triggered by either histone modifications or DNA methylation, both of which can result from targeting by endogenous siRNAs [11, 12]. Most of the endogenous siRNAs in plants match to repetitive sequences such as transposons, retrotransposons and centromeric repeats that are often heterochromatic [13]. siRNAs may also act as guides to target and degrade complementary mRNA molecules [14]. More detailed studies in several organisms continue to identify additional classes of small RNAs. In plants, this includes *trans*-acting siRNAs (ta-siRNAs) and natural-antisense siRNAs (nat-siRNAs), two relatively small classes of small RNAs that, due to space restrictions, we will not discuss here.

miRNA molecules originate from distinct genomic loci that encode RNAs predicted to form 'hairpin' structures that often have an imperfect double-stranded characteristic [9]. They are cleaved from the hairpin as a duplex by a DICER-LIKE1 protein (DCL1) and the miRNA strand of this duplex becomes associated with an AGO protein in RISC [15, 16]. In plants, miRNAs almost always induce cleavage and accelerated degradation of their 'target' mRNAs by forming base-pairing interactions [17, 18]. They can also direct cleavage of non-coding RNAs to induce production of ta-siRNAs [19, 20]. In addition, some miRNAs act by preventing mRNA translation and thereby limiting protein production [21, 22]. In contrast to siRNAs, miRNAs usually do not match perfectly to their target mRNA molecules [23–25].

In *Arabidopsis*, several genes, including the four members of the *DCL* family, have well-characterized roles in the biogenesis or function of small RNAs [26]. DCL1 is responsible for the synthesis of the majority of miRNAs [24, 27, 28], DCL2 and DCL4 facilitate the production of virus-derived siRNAs [29], and DCL3 is involved in the biosynthesis of endogenous siRNAs primarily from transposons and other repeats [10]. In addition, molecular and developmental phenotypes indicate functional redundancies amongst *DCL* genes in *Arabidopsis* [26]. RNA-dependent RNA polymerases (RdRPs or '*RDR*' genes) are also critical for siRNA biogenesis (six are encoded in the *Arabidopsis* genome), likely acting to produce dsRNAs as substrates for DICER-LIKE enzymes [30]. *rdr2* mutants show a dramatic reduction in endogenous siRNAs [10, 13], and *rdr6* mutants reduce siRNAs corresponding to certain viruses and transgenes, as well as affect the abundance of *trans*-acting siRNAs [19, 20, 31, 32]. DCL4 also plays a key role in producing endogenous RDR6-dependent ta-siRNAs [33]. Numerous genes important for small RNA biogenesis have been described and continue to be identified by genetic and biochemical approaches, but due to space limitations, we cannot describe these in detail. Interestingly, DNA methylation mutants also affect the abundance of many endogenous siRNAs. For instance, mutants in the de novo methyltransferase DRM2 cause a loss of siRNAs at some methylated loci such as *AtSN1* and the 5S rDNA [34], and *ddm1* and *met1* mutants cause a loss of some siRNAs corresponding to the knob on chromosome 4 [35]. These data suggest that the state of chromatin in part regulates small RNAs emanating from the underlying DNA sequences.

Deep Sequencing Analyses of Plant Small RNAs

Analyses using high-throughput, parallel sequencing of small RNAs have demonstrated that known and new miRNAs are among the most abundant plant small RNAs, whereas more than half of *Arabidopsis* small RNAs represent lower abundance siRNAs that match repetitive sequences, intergenic regions, and genes [13, 36, 37]. In these studies, a small number of miRNAs were cloned many thousands of times, whereas siRNAs tended to distribute across larger loci, consistent in both cases with their biogenesis. Centromeric regions have a high abundance of siRNAs [38], but small RNAs are distributed across all of the chromosomes. Deep sequencing data have also indicated that small RNA-based regulation in plants is highly specific and far more complex than previously demonstrated [13]. The *rdr2, dcl2/3/4* and *RNA pol IV* mutant lines are depleted for most heterochromatic siRNAs, and therefore greatly enriched for miRNAs and also for ta-siRNAs, although *dcl2/3/4* also lacks ta-siRNAs [39, 40]. For example, from *rdr2*, we were able to identify 13 new miRNAs that are absent from other plant genomes (*Medicago*, rice, and other plant sequences in GenBank). The use of such mutants is likely to be even more beneficial in other plant species, particularly those such as most crop plants that have larger genomes that have a deeper pool of siRNAs (see below).

Small RNA MPSS libraries from rice demonstrate that the complexity of small RNAs is greater in the larger rice genome than that of *Arabidopsis*, and rice has a high level of small RNA matches that are spread across the chromosomes rather than being concentrated around the centromeric region [41]. In addition, small RNAs that match the genome numerous times are much higher in rice; more than 260 small RNA sequences match the genome more than 1000-fold in rice, whereas no *Arabidopsis* sequences match at this level of redundancy [41]. These highly duplicitous small RNAs in rice are due to their origin in repeats, with a higher level of repeated sequences in rice compared to *Arabidopsis*. However, the higher abundance and higher number of distinct sequences in inflorescence versus seedlings are evident in both plant species, as are known miRNAs and many examples of regulated clusters of small RNAs. Comparisons across *Arabidopsis* and rice have identified numerous conserved small RNAs which are primarily miRNAs. The discovery of new miRNAs with unique characteristics, as well as more detailed studies of known miRNAs will likely lead to new computational tests to enable miRNA discovery; examples of this have been published [42, 43].

Conservation and Evolution of Plant Small RNAs

A substantial number of plant miRNAs are represented by numerous genomic loci. These miRNAs are given the same name but with different suffixes; for example, there are 14 copies of miR169 in the *Arabidopsis* genome (miR169a to miR169n). Within each miRNA family, different members are often closely related and sometimes give rise to identical miRNAs (fig. 1). These loci are believed to have a common origin, although there may be little sequence homology in the stem-loop. A comparison of *Arabidopsis* miRNA gene family members reveals a typical pattern of conservation. As shown in figure 1, high levels of conservation are found in the regions encoding the miRNA and miRNA*. In many cases, a lower level of conservation can be observed in the loop region of the secondary structure formed by the miRNA precursors. Because of the miRNA identity, it is a challenge to determine which loci from a particular miRNA family are expressed, although this can be done with promoter-reporter fusions. One recent study has indicated that almost all the miRNA genes in *Arabidopsis* are expressed [44]. miRNAs designated by different numbers may still share substantial homology, perhaps as a result of a common origin, and these are known as a miRNA family. In *Arabidopsis*, most of these multi-copy miRNAs are present as they are in other plant species and are easily detected by RNA gel blots or sequencing. Twenty-one miRNA gene families conserved in more than one plant species, representing 92 miRNA genes,

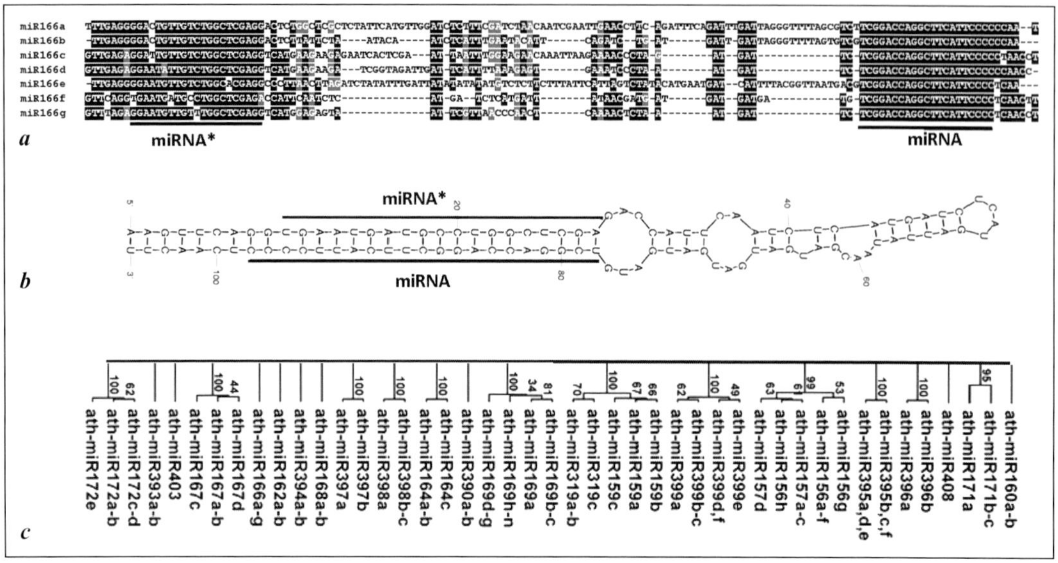

Fig. 1. miRNAs in *Arabidopsis*. ***a*** Alignment of the DNA sequence encoding the stem-loop structure for the miR166 family members. The conservation of the miRNA and miRNA* is apparent, as is the lower level of conservation of the sequences between them. ***b***The stem-loop structure for miR166f. The 5′ and 3′ ends of the primary miRNA transcript have been removed to focus on the hairpin. The miRNA-miRNA* duplex is cleaved by DCL1 to produce the characteristic dsRNA with a two nucleotide 3′ overhang. ***c*** A neighbor-joining phylogenetic tree of the *Arabidopsis* miRNAs which are conserved in other plant species. Not shown are more than 40 non-conserved *Arabidopsis* miRNAs. For miRNAs that are 100% identical in the final mature form, these were grouped together and are indicated in the names. For example, 'miR166a-g' represents seven identical family members, while 'miR399d,f' represents two separate loci (miR399d and miR399f). Numbers within the tree indicate bootstrap support, determined based on 1000 replicates.

have been identified in *Arabidopsis*. Nearly all members of these families are encoded by multiple genes (fig. 1c).

Cloning of miRNAs from moss and other lower land plants indicates that a small number of miRNAs are conserved over more than 400 million years of evolution [45–47]. While these data suggest that many miRNAs are evolutionarily conserved, some miRNAs are clearly not well conserved across species. Indeed, in *Arabidopsis*, extensive cloning and sequencing efforts have suggested that although the discovery of conserved miRNA genes reached the plateau, many more non-conserved miRNAs remain to be identified. These data suggest that a substantial set of miRNAs in a plant species is likely to be lineage-specific. In these datasets, more than 40 new non-conserved miRNAs have been identified [36, 39]. Nearly all of the non-conserved miRNAs are found at low expression levels and have eluded detection because of these low abundance levels. Scant data exist for understanding the evolution of plant miRNAs. However, two *Arabidopsis* miRNAs have been described as having evolved recently from duplicated gene fragments in which the small RNAs were originally produced as siRNAs [48]. Analyses of the similarity between the arms of miRNA precursors suggests that miRNA loci may evolve via inverted duplication events of protein-coding genes, and miRNA families may expand by duplication and divergence [49]. Additional studies are needed to determine the generality of this model, the rate of miRNA evolution and the proportion of miRNAs specific to individual lineages of plants.

Importance and Conservation of miRNA Targets

Identifying miRNA targets is the most significant step toward understanding the functional and biological significance of miRNA-mediated regulation. For miRNA families highly conserved across species, complementary sites in the target mRNAs are often also conserved (see the extensive list in [16]). Computational approaches based on genome-wide comparisons and sequence alignments can be used for predicting targets, with scoring based on experimentally-determined rules of miRNA-target interactions. Several miRNA target prediction tools are available for plants, one of which we have implemented on our website [25]. Unfortunately, this approach is only predictive and only applicable to species with significant genomic data. It is also not yet clear if the rules of miRNA-target interactions have been fully elucidated. Individual predicted targets are most often validated using 5′ RACE to map the 5′ end of the cleaved mRNA target to the middle of the miRNA sequence. Experimental identification of targets has advanced slowly and gene-by-gene; large-scale and parallel approaches would provide an even greater resource for researchers to predict their function, but have not yet been developed.

While considerable experimental work remains to be performed on the targets of both known and newly discovered miRNAs, it is clear that miRNAs are involved in a broad range of plant biological functions. Due to limited space and the large number of known miRNAs, it is impossible for us to summarize the functions of all known miRNAs, even for a single genome like that of *Arabidopsis*. A recent review by Mallory and Vaucheret has done this [50]. The biological roles of miRNAs are predominantly associated with development [4, 51, 52], but they also play a role in stress responses. Differential accumulation of miRNAs in different tissues is common and many miRNAs target transcription factor mRNAs [53, 54]. Moreover, developmental defects are associated with the miRNA metabolism mutants discussed above [19, 27, 55–58]. Although the association of small RNAs with viral resistance has been known for some time, more recent data demonstrate a role of miRNAs in plant disease resistance against bacterial or other pathogens [39, 59]. Associations of miRNAs and natural antisense siRNAs with abiotic and recently biotic stresses are also well-documented [60, 61]. In contrast to miRNAs, siRNA functions are broad, reflected also in the much broader set of siRNA-generating loci. siRNA functions include protection against viruses and mobile genetic elements, and other repetitive sequences [12, 30]. In addition, several genes are regulated at the level of chromatin modifications and DNA methylation, and small RNAs play a major role in the establishment and maintenance of these marks [62, 63], a pattern that is supported by a recent genome-wide study [64].

Conclusions

The first small RNAs were only discovered 14 years ago [65, 66] and were not known to be of general significance until about seven years ago [1, 67, 68]. Yet, they are now considered to be a major gene regulatory force in multicellular and some unicellular eukaryotes. Our work with *Arabidopsis* [13] and rice [41] indicates that small RNA populations in plants are so vast and complex that deep sequencing is necessary to elucidate the identity, regulation, and function of these interesting molecules. Projects to develop datasets of small RNA sequences derived from a broad range of plant species, many of which have few existing sequence data, will help to identify the sources and targets of small RNAs, to identify small RNAs unique to specific lineages or conserved more broadly across the sampled plant species. Such experiments would complement sequencing and annotation efforts by supplying data both on heterochromatic or repetitive regions and on novel and conserved miRNAs. This will be critical information for determining the function and activity of miRNAs and their target genes. Deep sequencing of small RNAs from *Arabidopsis* and rice has already identified hundreds of thousands of different (distinct) sequences matching to thousands of genomic loci. Many of these small RNAs are consistent with silencing of repetitive elements by siRNAs, although these data include matches to genes and intergenic regions consistent with targets and sources of miRNAs. The small size and exquisite specificity of small RNAs makes them ideal signaling molecules, and the potential for novel discoveries in this area of research is rich. Additional experiments focused on specific miRNAs are needed to clarify their targets, expression, and mode of action, while ongoing analyses of small RNA biogenesis pathways are likely to identify novel players in small RNA biochemistry.

Acknowledgements

This work was supported primarily by NSF grants 0439186 and 0548569.

References

1 Hamilton AJ, Baulcombe DC: A species of small antisense RNA in posttranscriptional gene silencing in plants. Science 1999;286:950–952.

2 Bartel B, Bartel DP: MicroRNAs: at the root of plant development? Plant Physiol 2003;132:709–717.

3 Mallory AC, Vaucheret H: MicroRNAs: something important between the genes. Curr Opin Plant Biol 2004;7:120–125.

4 Carrington JC, Ambros V: Role of microRNAs in plant and animal development. Science 2003;301:336–338.

5 Grishok A, Pasquinelli AE, Conte D, Li N, Parrish S, et al: Genes and mechanisms related to RNA interference regulate expression of the small temporal RNAs that control *C. elegans* developmental timing. Cell 2001;106:23–34.

6 Ketting RF, Fischer SE, Bernstein E, Sijen T, Hannon GJ, Plasterk RH: Dicer functions in RNA interference and in synthesis of small RNA involved in developmental timing in *C. elegans*. Genes Dev 2001;15:2654–2659.

7 Hutvagner G, McLachlan J, Pasquinelli AE, Balint E, Tuschl T, Zamore PD: A cellular function for the RNA-interference enzyme Dicer in the maturation of the let-7 small temporal RNA. Science 2001;293:834–838.

8 Lee YS, Nakahara K, Pham JW, Kim K, He Z, Sontheimer EJ, Carthew RW: Distinct roles for *Drosophila* Dicer-1 and Dicer-2 in the siRNA/miRNA silencing pathways. Cell 2004;117:69–81.

9 Bartel DP: MicroRNAs: genomics, biogenesis, mechanism, and function. Cell 2004;116:281–297.

10 Xie Z, Johansen LK, Gustafson AM, Kasschau KD, Lellis AD, et al: Genetic and functional diversification of small RNA pathways in plants. PLoS Biol 2004;2:E104.

11 Matzke MA, Matzke AJ, Pruss GJ, Vance VB: RNA-based silencing strategies in plants. Curr Opin Genet Dev 2001;11:221–227.

12 Lippman Z, Martienssen R: The role of RNA interference in heterochromatic silencing. Nature 2004;431:364–370.

13 Lu C, Tej SS, Luo S, Haudenschild CD, Meyers BC, Green PJ: Elucidation of the small RNA component of the transcriptome. Science 2005;309:1567–1569.

14 Hannon GJ: RNA interference. Nature 2002;418:244–251.

15 Schwarz DS, Hutvagner G, Haley B, Zamore PD: Evidence that siRNAs function as guides, not primers, in the *Drosophila* and human RNAi pathways. Mol Cell 2002;10:537–548.

16 Jones-Rhoades MW, Bartel DP, Bartel B: MicroRNAs and their regulatory roles in plants. Annu Rev Plant Biol 2006;57:19–53.

17 Tang G, Reinhart BJ, Bartel DP, Zamore PD: A biochemical framework for RNA silencing in plants. Genes Dev 2003;17:49–63.

18 Llave C, Xie Z, Kasschau KD, Carrington JC: Cleavage of Scarecrow-like mRNA targets directed by a class of *Arabidopsis* miRNA. Science 2002;297:2053–2056.

19 Peragine A, Yoshikawa M, Wu G, Albrecht HL, Poethig RS: SGS3 and SGS2/SDE1/RDR6 are required for juvenile development and the production of trans-acting siRNAs in *Arabidopsis*. Genes Dev 2004;18:2368–2379.

20 Vazquez F, Vaucheret H, Rajagopalan R, Lepers C, Gasciolli V, et al: Endogenous trans-acting siRNAs regulate the accumulation of *Arabidopsis* mRNAs. Mol Cell 2004;16:69–79.
21 Aukerman MJ, Sakai H: Regulation of flowering time and floral organ identity by a MicroRNA and its APETALA2-like target genes. Plant Cell 2003;15:2730–2741.
22 Chen X: A microRNA as a translational repressor of APETALA2 in *Arabidopsis* flower development. Science 2004;303:2022–2025.
23 Kasschau KD, Xie Z, Allen E, Llave C, Chapman EJ, Krizan KA, Carrington JC: P1/HC-Pro, a viral suppressor of RNA silencing, interferes with *Arabidopsis* development and miRNA function. Dev Cell 2003;4:205–217.
24 Reinhart BJ, Weinstein EG, Rhoades MW, Bartel B, Bartel DP: MicroRNAs in plants. Genes Dev 2002;16:1616–1626.
25 Jones-Rhoades MW, Bartel DP: Computational identification of plant microRNAs and their targets, including a stress-induced miRNA. Mol Cell 2004;14:787–799.
26 Henderson IR, Zhang X, Lu C, Johnson L, Meyers BC, Green PJ, Jacobsen SE: Dissecting *Arabidopsis thaliana* DICER function in small RNA processing, gene silencing and DNA methylation patterning. Nat Genet 2006;38:721–725.
27 Park W, Li J, Song R, Messing J, Chen X: CARPEL FACTORY, a Dicer homolog, and HEN1, a novel protein, act in microRNA metabolism in *Arabidopsis thaliana*. Curr Biol 2002;12:1484–1495.
28 Kurihara Y, Watanabe Y: *Arabidopsis* micro-RNA biogenesis through Dicer-like 1 protein functions. Proc Natl Acad Sci USA 2004;101:12753–12758.
29 Deleris A, Gallego-Bartolome J, Bao J, Kasschau KD, Carrington JC, Voinnet O: Hierarchical action and inhibition of plant Dicer-like proteins in antiviral defense. Science 2006;313:68–71.
30 Baulcombe D: RNA silencing in plants. Nature 2004;431:356–363.
31 Dalmay T, Hamilton A, Rudd S, Angell S, Baulcombe DC: An RNA-dependent RNA polymerase gene in *Arabidopsis* is required for posttranscriptional gene silencing mediated by a transgene but not by a virus. Cell 2000;101:543–553.
32 Mourrain P, Beclin C, Elmayan T, Feuerbach F, Godon C, et al: *Arabidopsis SGS2* and *SGS3* genes are required for posttranscriptional gene silencing and natural virus resistance. Cell 2000;101: 533–542.
33 Gasciolli V, Mallory A, Bartel D, Vaucheret H: Partially redundant functions of *Arabidopsis* DICER-like enzymes and a role for DCL4 in producing trans-acting siRNAs. Curr Biol 2005; 15:1–7.
34 Zilberman D, Cao X, Johansen LK, Xie Z, Carrington JC, Jacobsen SE: Role of *Arabidopsis* ARGONAUTE4 in RNA-directed DNA methylation triggered by inverted repeats. Curr Biol 2004;14:1214–1220.
35 Lippman Z, May B, Yordan C, Singer T, Martienssen R: Distinct Mechanisms Determine transposon inheritance and methylation via small interfering RNA and histone modification. PLoS Biol 2003;1:E67.
36 Rajagopalan R, Vaucheret H, Trejo J, Bartel DP: A diverse and evolutionarily fluid set of microRNAs in *Arabidopsis thaliana*. Genes Dev 2006;20:3407–3425.
37 Kasschau KD, Fahlgren N, Chapman EJ, Sullivan CM, Cumbie JS, Givan SA, Carrington JC: Genome-wide profiling and analysis of *Arabidopsis* siRNAs. PLoS Biol 2007;5:e57.
38 Yan H, Ito H, Nobuta K, Ouyang S, Jin W, et al: Genomic and genetic characterization of rice Cen3 reveals extensive transcription and evolutionary implications of a complex centromere. Plant Cell 2006;18:2123–2133.
39 Lu C, Kulkarni K, Souret FF, MuthuValliappan R, Tej SS, et al: MicroRNAs and other small RNAs enriched in the *Arabidopsis* RNA-dependent RNA polymerase-2 mutant. Genome Res 2006;16: 1276–1288.
40 Zhang X, Henderson IR, Lu C, Green PJ, Jacobsen SE: Role of RNA polymerase IV in plant small RNA metabolism. Proc Natl Acad Sci USA 2007;104:4536–4541.
41 Nobuta K, Venu RC, Lu C, Belo A, Vemaraju K, et al: An expression atlas of rice mRNA and small RNA. Nat Biotechnol 2007;25:473–477.
42 Bonnet E, Wuyts J, Rouze P, Van de Peer Y: Evidence that microRNA precursors, unlike other non-coding RNAs, have lower folding free energies than random sequences. Bioinformatics 2004;20: 2911–2917.

43 Ohler U, Yekta S, Lim LP, Bartel DP, Burge CB: Patterns of flanking sequence conservation and a characteristic upstream motif for microRNA gene identification. RNA 2004;10:1309–1322.
44 Xie Z, Allen E, Fahlgren N, Calamar A, Givan SA, Carrington JC: Expression of *Arabidopsis* MIRNA Genes. Plant Physiol 2005;138:2145–2154.
45 Floyd SK, Bowman JL: Gene regulation: ancient microRNA target sequences in plants. Nature 2004;428:485–486.
46 Arazi T, Talmor-Neiman M, Stav R, Riese M, Huijser P, Baulcombe DC: Cloning and characterization of micro-RNAs from moss. Plant J 2005;43:837–848.
47 Axtell MJ, Bartel DP: Antiquity of microRNAs and their targets in land plants. Plant Cell 2005;17:1658–1673.
48 Allen E, Xie Z, Gustafson AM, Sung GH, Spatafora JW, Carrington JC: Evolution of microRNA genes by inverted duplication of target gene sequences in *Arabidopsis thaliana*. Nat Genet 2004;36:1282–1290.
49 Fahlgren N, Howell MD, Kasschau KD, Chapman EJ, Sullivan CM, et al: High-throughput sequencing of *Arabidopsis* microRNAs: Evidence for frequent birth and death of MIRNA genes. PLoS ONE 2007;2:e219.
50 Mallory AC, Vaucheret H: Functions of microRNAs and related small RNAs in plants. Nat Genet 2006;38(suppl):S31–S36.
51 Palatnik JF, Allen E, Wu X, Schommer C, Schwab R, Carrington JC, Weigel D: Control of leaf morphogenesis by microRNAs. Nature 2003;425:257–263.
52 Kidner CA, Martienssen RA: The developmental role of microRNA in plants. Curr Opin Plant Biol 2005;8:38–44.
53 Rhoades MW, Reinhart BJ, Lim LP, Burge CB, Bartel B, Bartel DP: Prediction of plant microRNA targets. Cell 2002;110:513–520.
54 Llave C, Kasschau KD, Rector MA, Carrington JC: Endogenous and silencing-associated small RNAs in plants. Plant Cell 2002;14:1605–1619.
55 Vazquez F, Gasciolli V, Crete P, Vaucheret H: The nuclear dsRNA binding protein HYL1 is required for microRNA accumulation and plant development, but not posttranscriptional transgene silencing. Curr Biol 2004;14:346–351.
56 Jacobsen SE, Running MP, Meyerowitz EM: Disruption of an RNA helicase/RNAse III gene in *Arabidopsis* causes unregulated cell division in floral meristems. Development 1999;126: 5231–5243.
57 Vaucheret H, Vazquez F, Crete P, Bartel DP: The action of ARGONAUTE1 in the miRNA pathway and its regulation by the miRNA pathway are crucial for plant development. Genes Dev 2004;18:1187–1197.
58 Kidner CA, Martienssen RA: Spatially restricted microRNA directs leaf polarity through ARGONAUTE1. Nature 2004;428:81–84.
59 Navarro L, Dunoyer P, Jay F, Arnold B, Dharmasiri N, et al: A plant miRNA contributes to antibacterial resistance by repressing auxin signaling. Science 2006;312: 436–439.
60 Borsani O, Zhu J, Verslues PE, Sunkar R, Zhu JK: Endogenous siRNAs derived from a pair of natural cis-antisense transcripts regulate salt tolerance in *Arabidopsis*. Cell 2005;123: 1279–1291.
61 Sunkar R, Zhu JK: Novel and stress-regulated microRNAs and other small RNAs from *Arabidopsis*. Plant Cell 2004;16:2001–2019.
62 Chan SW, Zilberman D, Xie Z, Johansen LK, Carrington JC, Jacobsen SE: RNA silencing genes control de novo DNA methylation. Science 2004;303:1336.
63 Kinoshita T, Miura A, Choi Y, Kinoshita Y, Cao X, et al: One-way control of FWA imprinting in *Arabidopsis* endosperm by DNA methylation. Science 2004;303:521–523.
64 Zhang X, Yazaki J, Sundaresan A, Cokus S, Chan SW, et al: Genome-wide high-resolution mapping and functional analysis of DNA methylation in *Arabidopsis*. Cell 2006;126:1189–1201.
65 Lee RC, Feinbaum RL, Ambros V: The *C. elegans* heterochronic gene lin-4 encodes small RNAs with antisense complementarity to lin-14. Cell 1993;75:843–854.
66 Wightman B, Ha I, Ruvkun G: Posttranscriptional regulation of the heterochronic gene lin-14 by lin-4 mediates temporal pattern formation in *C. elegans*. Cell 1993;75:855–862.

67 Fire A, Xu S, Montgomery MK, Kostas SA, Driver SE, Mello CC: Potent and specific genetic interference by double-stranded RNA in *Caenorhabditis elegans*. Nature 1998;391:806.
68 Reinhart BJ, Slack FJ, Basson M, Pasquinelli AE, Bettinger JC, et al: The 21-nucleotide let-7 RNA regulates developmental timing in *Caenorhabditis elegans*. Nature 2000;403:901–906.

Blake C. Meyers
Department of Plant and Soil Sciences and Delaware Biotechnology Institute
Newark, DE 19711 (USA)
Tel. +1 302 831 3418, Fax +1 302 831 4841, E-Mail meyers@dbi.udel.edu

Volff J-N (ed): Plant Genomes.
Genome Dyn. Basel, Karger, 2008, vol 4, pp 119–130

Recent Insights Into the Evolution of Genetic Diversity of Maize

A. Rafalski, S. Tingey

DuPont Crop Genetics Research, DuPont Experimental Station, Wilmington, Del., USA

Abstract

Evolutionary changes that occur within the maize genome can be divided into two classes: In the protein-coding regions, mutations that survive selective pressure primarily consist of single nucleotide polymorphisms which evolve at a relatively slow rate (10^{-9} mutations/bp/generation), and functionally detrimental insertions are strongly selected against. In intergenic regions rapidly evolving ($10^{-4} \sim 10^{-8}$ mutations/bp/generation) transposon insertions and deletions predominate. While genic single nucleotide changes are expected in part to result in amino acid sequence variants leading to the modification of protein properties (catalytic properties of enzymes, binding constants, etc.), transposable elements and other large insertions and deletions, when functionally relevant, are predicted to be regulatory in nature and may affect gene expression at distances of up to 100 kb away. Here, we discuss recent experimental evidence for massive dynamic changes of maize intergenic regions and the predicted functional consequences of genome diversity within this species.

Maize has been known as one of the most diverse species at both the phenotypic and at the genotypic level. McClintock's studies of mobile 'controlling elements' and their effects on gene expression were among the first to reveal a dynamic genome. We are just now beginning to understand some of her observations at the molecular level [1]. Until recently, most studies of DNA sequence diversity focused on the least variable component of the genome; the genes themselves. Only recently information has been generated to allow DNA sequence comparisons of intergenic segments within the maize genome. These studies have led to a better understanding of the types of changes that occur, the allelic diversity that might exist, the rate at which these changes occur and helped generate ideas about the functional relevance of such changes. In this chapter we are going to discuss the dynamic nature of the maize genome. Some of these results challenge our conventional understanding of intraspecific genetic variation suggesting possible new functions of intergenic 'junk' DNA. The evolution of maize genome sequence in the context of interspecific genetic variation will not be covered. For a recent article see [2].

Intraspecific Genetic Diversity and Species Identity

Interspecific genetic diversity is usually a subject of taxonomy while population genetics concerns itself with intraspecific diversity. At the molecular level, these distinctions are not always clear. Intraspecific DNA sequence diversity of maize is up to 10× higher than that of humans, and variation within genes can approach the levels of divergence found between humans and chimpanzees [3]. One may ask, for example, what level of DNA sequence diversity is compatible with the maintenance of species identity. As we will attempt to show, it appears that in maize, sequence organization and level of sequence conservation within intergenic regions is not stringently controlled. One hypothesis would be that little more than the colinearity of most, although not all genes, together with their regulatory regions, is all that is required for sexual compatibility and species identity.

Studies of Genic and Intergenic Diversity

Early studies of genetic diversity focused on DNA sequence divergence within genes. This was due to an interest in gene evolution, and also due to the technical difficulties in the analysis of repetitive sequences which comprise the majority of intergenic segments of the maize genome and cannot easily be accessed through PCR technology. The overall features of the organization of intergenic regions have been elucidated (fig. 1) [4–6]. Only recently, however, comparative sequencing of larger allelic regions from multiple inbred lines has been undertaken [7–9]. The major insight from these studies is that the nature of observed allelic differences is very different within genes and their immediate vicinity from those observed within intergenic regions. The maize genome has undergone massive changes within the intergenic regions, with examples of less than 50% DNA sequence conservation, and these changes are most likely continuing. This is in contrast to species such as humans the genomes of which, after period of rapid change, appear to be in a quiescent phase characterized by relatively low intraspecific diversity.

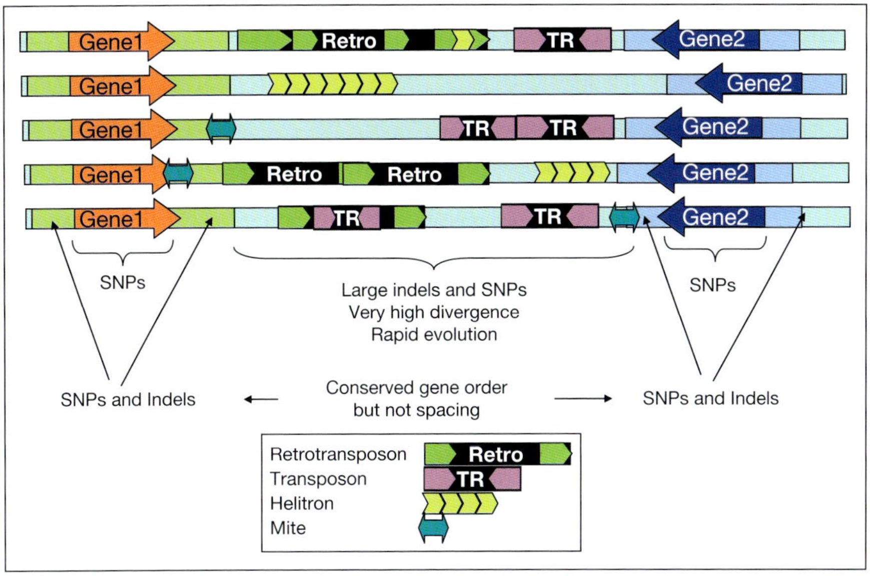

Fig. 1. Organization of the maize genome. Five different alleles of a short segment encompassing two genes are represented schematically.

Gene Diversity of Maize

The domestication of maize from its ancestor teosinte, and the genetic diversity within the *Zea* genus has been the subject of many studies. All available methodologies including restriction fragment length polymorphisms (RFLP), simple sequence repeats (SSRs), more recently, single nucleotide polymorphisms (SNPs) have been applied as estimators of genetic divergence [10–17]. We have been especially interested in the genetic diversity that exists within cultivated maize, including elite breeding material. To this end, we have focused on the widely available maize expressed sequenced tag sequences. These sequences could be used to develop PCR amplicons comprising 3′-UTR and last exon sequences of 300–800 bp for analysis. Direct PCR sequencing from a collection of inbred lines allows the identification of SNPs and insertion/deletions (indels) at a locus [17]. Indels are very rare in the coding regions but occur frequently in 3′-UTRs. The SNP frequency is approximately 3-fold higher in the UTRs relative to exonic sequences. Across a representative collection of cultivated maize SNPs occur with an approximate frequency of 1 in 130 nt in exons, and 1 in 48 nt in the UTRs. The SNP diversity in elite maize was estimated to be $\pi = 0.0063$ [17] and is approximately ten fold higher than that in human populations

Table 1. Causes of genome change

Agent of change	Location	Frequency
Single nucleotide change	Anywhere	10^{-9}
Simple sequence repeats	Anywhere, may be more frequent near genes [15, 20]	10^{-4}
Indels	Anywhere, rarely observed in genes	Unknown
Gene duplication	Frequently as tandem repeats	Rare, unknown
Retrotransposons	Commonly into other retrotransposons	Very frequent
Transposons	Some near genes (Mu)	Frequent to rare
Transposons: Mites	Near genes?	Unknown
Transposons: Helitrons	Anywhere, slightly more common near telomeres	Frequent
Transposons: Pack-Mules	Anywhere, capture genes	Unknown

(π = 0.0007, http://egp.gs.washington.edu/summary_ stats.html), and three fold higher than in soybean (*Glycine max*, π = 0.00175) [18].

Many different mechanisms of change operate in the maize genome (table 1). While most of these agents of change operate across the entire genome, in genic regions, due to selection against loss of function mutations, we only observe viable changes. Those are most frequently silent single nucleotide mutations, occurring with a frequency in the range of 10^{-9} per nucleotide per generation [19], or functionally acceptable amino-acid changing mutations. Indels are more likely to disrupt protein structure and lead to an inactive gene. As a result, rare indels of 3 bp or a multiple of 3 bp are tolerated; other gene inactivating mutations are rapidly eliminated from the population. In contrast, indels are rather common in UTRs and introns. Simple sequence repeats are a special case of indel mutations and have been reported to occur more frequently in non-repetitive segments of the genome [20].

Intergenic Sequence Diversity in Maize

There still is a paucity of detailed data describing intraspecific diversity in intergenic regions especially for inbreds other than B73 and Mo17. The first comparative sequences from the Bz locus [7] found little intergenic sequence similarity between B73 and Mo17 [7, 21]. Other recently sequenced maize inbreds are equally diverse in the intergenic regions examined [22]. The presence of many retroelement insertions was not suprising; the involvement of retroelements in the organization of maize intergenic regions has been well documented [5, 23, 24]. However, the enormous allelic diversity in these regions was unexpected. It indicated that the genome of maize has been under an assault from various mobile genetic elements in the relatively recent past

(<1,000,000 years), and that fixation or elimination of new mobile genetic element insertions has not occurred for the majority of such events. In fact, in the regions sequenced, on average 50% of the intergenic sequences are not homologous in the strict sense in that their origins are different [9]. Initial reports indicated significant differences in the content of genes between the B73 and Mo17 at the Bz locus [7]. Subsequent research demonstrated that these differences are in pseudogenes [25]. Genic differences (indels) have been found at other loci [8, 21]. Other examples of extremely high allelic variation have recently been described in maize [2, 9, 25, 26] and in barley [27].

Brunner et al. sequenced four loci 126–406 kbp in size from the two maize inbreds B73 and Mo17 [9]. Between 21% (Adh1, chromosome 1L, bin 1.10) and 68.8% (locus 9002, chromosome 1L bin 1.08) of the sequences were different between the two alleles providing a generalization of the previously reported observations [7]. Most of the differences between alleles were accounted for by recently (<1 my) inserted retrotransposons. Genes were conserved between alleles and were largely co-linear with rice [9]. Intergenic sequence segments with very high homology to genes turned out to be *Helitron* transposons carrying gene fragments (pseudogenes, see below) [25, 26].

Sources and Significance of Sequence Diversity in Different Regions of the Genome

Much of the evolutionary theory has been developed with the focus on evolution of the coding regions of genes, and attention has been paid to the silent vs. non-silent mutations and their role in the evolution of protein function [28, 29]. Excellent examples of rapid changes include the evolution of disease resistance genes, although best examples are from species other than maize (for an example, [30]). In genic regions, gene duplications are frequently followed by sequence and possible functional diversification [8].

It has been recognized that mutations in regulatory sequences may be just as important as genic mutations, as illustrated by the introduction of morphological innovations [31, 32]. Regulatory mutations are likely to occur at the sites of the binding of transcription factors frequently found in promoter or 5′-UTR regions [31]. Such mutations may occur at relatively distant sites, perhaps as far away as ten times the length of the gene [33]. These results underscore the functional importance of what is generally thought of as intergenic 'junk' DNA, sequences that in maize are also the most variable.

Intergenic regions of maize are full of retroelements, transposons, tandem repeats, SNPs and other DNA sequence changes [2, 9, 23, 34]. High rates of transposon activity that create dramatic genome changes are likely to be functionally important, although specific information is lacking in most cases. In *Oryza australiensis* several distinct explosions of retroelement insertions over a relatively short time, the most recent of them occurring ~200,000 years ago, which led to a doubling of the genome [35], suggest a very high rate of mutation during the period of peak activity.

It is common to find that the immediate environment of an active gene is divergent between different maize inbred lines [8, 9]. It is reasonable to hypothesize that such changes in the genic environment will affect gene regulation, for example by affecting integrity of promoters or other regulatory elements, providing new regulatory sequences, causing antisense transcription or affecting chromatin structure. Experimental evidence for the effects of retroelement insertions on the expression of nearby genes is being collected [36].

In an organism like maize that evolved through a genome duplication, gene functions are frequently redundant allowing for gene loss. Alternatively, a change in expression pattern of the duplicated gene, caused by a retrotransposon insertion in the vicinity of the gene, may occasionally play an adaptive role.

Different alleles with different regulatory properties, frequently caused by transposon activity may coexist in the population [37, 38]. It is tempting to speculate that occasionally such regulatory variants of the same gene may be brought together in a hybrid, and contribute to the phenomenon of heterosis.

Genes

While most of the genes are conserved between diverse maize inbreds (genotypes), instances of gene duplication and deletion are relatively common. Song and collaborators [8] described the reorganization of a zein gene cluster in maize which resulted in substantial differences in the structure of this locus between maize inbreds but also in concomitant changes in the regulation of gene expression. Similar differences in gene complement have been noted before, for example in clusters of disease resistance genes [39]. In maize, disease resistance genes have also been observed in some cultivars that are completely absent from others (P. Wolters, unpublished observations). It would be of interest to determine the minimal set of maize genes required for maintenance of species identity.

Retroelements

Retroelements constitute 70–80% of cereal genomes [4, 5, 23] indicating that their insertions have been a major mechanism of genome expansion. Recombination between long terminal repeats of retroelements is one of the mechanisms of genome contraction in grass genomes [6, 40]. In maize, however, few solo LTRs occur indicating that frequency of retrotransposon elimination by recombination is low [9].

The insertion time of individual retroelement copies may be estimated from the divergence of the long terminal repeats [24]. Such studies indicate that retroelement insertions have been occurring at least over the last several millions of years in maize [35]. While LTR sequence divergence does not allow a precise estimate of the age of more recently inserted elements, the persistence of both the element and of the empty insertion site in other individuals suggests a recent origin. Neither fixation nor elimination occurred via a genetic drift mechanism within the last 1–2 million years [9]. In fact, among the elements which have not yet become fixed in the maize gene pool, the youngest ones (<0.5 my) are most common [9]. In rice some retroelements, among them Tos 17, are induced by stress and continue to cause genome changes [41].

Transposons

Transposons are among the best documented agents of genome change in maize, and many reviews exist. Therefore we do not devote much time here to these elements. We will mention recently described DNA transposons which carry gene fragments and therefore may contribute to the rapid evolution of gene content and function. The Wessler group identified a group of small elements named Pack-Mules which appear to acquire gene fragments, and less frequently whole genes and transpose them to different locations in the genome [42]. Pack-Mules appear to insert close to genes (although a statistically rigorous analysis is lacking) and thus may acquire genes in the process of excision between two elements flanking a gene. In rice, a large number of genes appear to have been translocated by Pack-Mules [42, 43].

A more recent addition to a large repertoire of maize transposons is a *Helitron* element [44]. These transposons were first noticed by computational analysis in several species including *Arabidopsis thaliana* [45], and subsequently been described in maize [46].

More detailed analysis identified *Helitrons* by comparative sequencing of intergenic regions in maize and revealed that many of them contain multiple gene fragments derived from distant loci in the genome [21, 25, 46]. The history of one such family seems to indicate that in the process of replication *Helitrons* acquire and lose gene segments over time through an unknown mechanism [26].

Unlike retroelements, the insertion time of *Helitrons* cannot be determined easily but an analysis of a small family of *Helitrons* in maize allowed Brunner et al. [26] to propose their relationship to each other and order of insertion. The fact that many *Helitrons* are not shared between different alleles indicates that they most likely inserted over the last couple of million years. As expected, *Helitrons* contained in present day maize originate pre-domestication although their frequencies vary widely between maize and teosinte (fig. 2).

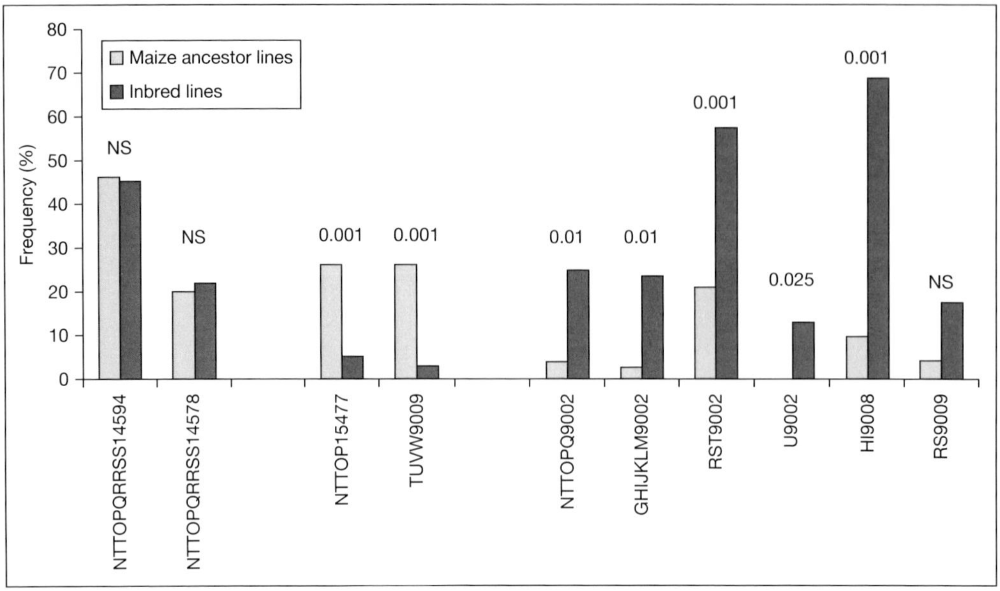

Fig. 2. Allele frequencies of several *Helitron* elements in maize and teosinte (S. Brunner, unpublished data). Statistical significance of the differences between the frequency of *Helitrons* in maize and teosinte is indicated on the bar graph. NS = non significant.

Brunner et al. [26] demonstrated that some of the *Helitron* elements are transcribed and the resulting RNA is spliced. Multiple gene fragments, including fragments of different genes, are concatenated and at times an open reading frame is maintained in the spliced transcript. This observation suggests that over evolutionary time *Helitrons* may contribute to the evolution of novel gene functions by combining different functional domains from different genes. This phenomenon has not been demonstrated yet in maize, but such a mechanism of gene evolution was proposed by Gilbert [47] and has been demonstrated in a few cases [48, 49]. Even though most of the *Helitron*-derived transcripts may be defective, they may also participate in epigenetic regulation through various RNAi pathways.

Tandem Repeats

Maize like other species contains many different types of tandemly repeated sequences including telomeric and centromeric repeats. Recently, repeat content and chromosomal localization of tandem repeats in different maize inbreds has been compared [50, and E. Ananiev, personal communication]. Chromosome painting of F_1 hybrids (Mo17 × B17) demonstrated large differences between the two lines in both size and location of major tandem repeat clusters (E. Ananiev, personal communication). Very large segments of microsatellite-like tandem repeats called 'megatracts' have been discovered in the maize genome [51]. These megatracts like other repeat classes show extreme allelic variability in the maize genome.

Lack of Intergenic Colinearity and Recombination

As discussed above, the continuing activity of numerous types of mobile elements creates large segments of intergenic space in maize that has no allelic counterpart in other lines. As a result, in certain crosses recombination is altered in some regions because of lack of homology. This effect may contribute to the phenomenon of linkage drag, or may influence how frequently allelic variation can be recombined. In addition, even within regions of very good DNA sequence homology, recombination in maize occurs primarily in genes [52]. Recent data demonstrates that the recombination frequency in maize is indeed strongly correlated with gene density [53, K. Fengler, personal communication].

Conclusions

In highly diverse species such as maize, a genome sequence of a single accession will be inadequate to describe the full content of regulatory sequences, and to a lesser extent, the genes. The maize genome sequencing project is focused on a BAC-by-BAC sequencing approach for the inbred variety B73, and a shotgun sequencing approach is being taken for chromosome 10 from the inbred variety Mo17 (http://www.maizegdb.org/sequencing_project.php). Until sequencing technology enables rapid and inexpensive re-sequencing of large genomes, genetic diversity studies targeted to specific loci and chromosome regions could provide much needed information about structural and functional diversity of intergenic sequences including their role in genetic and epigenetic regulation.

Acknowledgements

We would like to thank Bailin Li and Michele Morgante for stimulating discussions, Evgueni Ananiev, Kevin Fengler and Stephan Brunner for sharing unpublished results.

References

1 Alleman M, Sidorenko L, McGinnis K, Seshadri V, Dorweiler JE, et al: An RNA-dependent RNA polymerase is required for paramutation in maize. Nature 2006;442:295–298.

2 Bruggmann R, Bharti A, Gundlach H, Lai J, Young S, et al: Uneven chromosome contraction and expansion in the maize genome. Genome Res 2006;16:1241–1251.

3 Buckler ES, Gaut BS, McMullen MD: Molecular and functional diversity of maize. Curr Opin Plant Biol 2006;9:172–176.

4 Meyers BC, Tingey SV, Morgante M: Abundance, distribution and transcriptional activity of repetitive elements in the maize genome. Genome Res 2001;11:1660–1676.

5 Kumar A, Bennetzen JL: Retrotransposons: central players in the structure, evolution and function of plant genomes. Trends Plant Sci 2000;5:509–510.

6 Bennetzen JL: Mechanisms and rates of genome expansion and contraction in flowering plants. Genetica 2002;115:29–36.

7 Fu H, Dooner HK: Intraspecific violation of genetic colinearity and its implications in maize. Proc Natl Acad Sci USA 2002;99:9573–9578.

8 Song R, Messing J: Gene expression of a gene family in maize based on noncollinear haplotypes. Proc Natl Acad Sci USA 2003;100:9055–9060.

9 Brunner S, Fengler K, Morgante M, Tingey S, Rafalski A: Evolution of DNA sequence nonhomologies among maize inbreds. Plant Cell 2005;17:343–360.

10 Shattuck-Eidens DM, Bell RN, Neuhausen SL, Helentjaris T: DNA sequence variation within maize and melon: observations from polymerase chain reaction amplification and direct sequencing. Genetics 1990;126:207–217.

11 Gaut BS, Doebley JF: DNA sequence evidence for the segmental allotetraploid origin of maize. Proc Natl Acad Sci USA 1997;94:6809–6814.

12 Wang R-L, Stec A, Hey J, Lukens L, Doebley J: The limits of selection during maize domestication. Nature 1999;398:236–239.

13 Tenaillon MI, Sawkins MC, Long AD, Gaut RL, Doebley JF, Gaut BS: Patterns of DNA sequence polymorphism along chromosome 1 of maize (*Zea mays* ssp. *mays* L.). Proc Natl Acad Sci USA 2001;98:9161–9166.

14 Matsuoka Y, Vigouroux Y, Goodman MM, Sanchez GJ, Buckler E, Doebley J: A single domestication for maize shown by multilocus microsatellite genotyping. Proc Natl Acad Sci USA 2002;99: 6080–6084.

15 Vigouroux Y, Jaqueth JS, Matsuoka Y, Smith OS, Beavis WD, Smith JS, Doebley J: Rate and pattern of mutation at microsatellite loci in maize. Mol Biol Evol 2002;19:1251–1260.

16 Clark RM, Linton E, Messing J, Doebley JF: Pattern of diversity in the genomic region near the maize domestication gene *tb1*. Proc Natl Acad Sci USA 2004;101:700–707.

17 Ching A, Caldwell KS, Jung M, Dolan M, Smith OS, et al: SNP frequency, haplotype structure and linkage disequilibrium in elite maize inbred lines. BMC Genet 2002;3:19.

18 Zhu YL, Song QJ, Hyten DL, Van Tassell CP, Matukumalli LK, et al: Single-nucleotide polymorphisms in soybean. Genetics 2003;163:1123–1134.

19 Gaut BS, Morton BR, McCaig BM, Clegg MT: Substitution rate comparisons between grasses and palms: synonymous rate differences at the nuclear gene *Adh1* parallel rate differences at the plastid gene rblL. Proc Natl Acad Sci USA 1996;93:10274–10279.

20 Morgante M, Hanafey M, Powell W: Microsatellites are preferentially associated with nonrepetitive DNA in plant genomes. Nat Genet 2002;30:194–200.

21 Lai J, Li Y, Messing J, Dooner HK: Gene movement by *Helitron* transposons contributes to the haplotype variability of maize. Proc Natl Acad Sci USA 2005;102:9068–9073.

22 Wang Q, Dooner HK: Remarkable variation in maize genome structure inferred from haplotype diversity at the *bz* locus. Proc Natl Acad Sci USA 2006;103:17644–17649.

23 SanMiguel P, Tikhonov A, Jin YK, Motchoulskaia N, Zakharov D, et al: Nested retrotransposons in the intergenic regions of the maize genome. Science 1996;274:765–768.

24 SanMiguel P, Gaut BS, Tikhonov A, Nakajima Y, Bennetzen JL: The paleontology of intergene retrotransposons of maize. Nat Genet 1998;20:43–45.

25 Morgante M, Brunner S, Pea G, Fengler K, Zuccolo A, Rafalski A: Gene duplication and exon shuffling by helitron-like transposons generate intraspecies diversity in maize. Nat Genet 2005;37: 997–1002.

26 Brunner S, Pea G, Rafalski A: Origins, genetic organization and transcription of a family of non-autonomous helitron elements in maize. Plant J 2005;43:799–810.

27 Scherrer B, Isidore E, Klein P, Kim JS, Bellec A, et al: Large intraspecific haplotype variability at the rph7 locus results from rapid and recent divergence in the barley genome. Plant Cell 2005; 17:361–374.

28 Hanson MA, Gaut BS, Stec AO, Fuerstenberg SI, Goodman MM, Coe EH, Doebley JF: Evolution of anthocyanin biosynthesis in maize kernels: the role of regulatory and enzymatic loci. Genetics 1996;143:1395–1407.

29 Gaut BS, Clegg MT: Nucleotide polymorphism in the *Adh1* locus of pearl millet (*Pennisetum glaucum*) (Poaceae). Genetics 1993;135:1091–1097.

30 Meyers BC, Shen KA, Rohani P, Gaut BS, Michelmore RW: Receptor-like genes in the major resistance locus of lettuce are subject to divergent selection. Plant Cell 1998;11:1833–1846.

31 Doebley J, Lukens L: Transcriptional regulators and the evolution of plant form. Plant Cell 1998; 10:1075–1082.

32 de Meaux J, Pop A, Mitchell-Olds T: Cis-regulatory evolution of Chalcone-synthase expression in the genus *Arabidopsis*. Genetics 2006;174:2181–2202.

33 Clark RM, Wagler TN, Quijada P, Doebley J: A distant upstream enhancer at the maize domestication gene *tb1* has pleiotropic effects on plant and inflorescent architecture. Nat Genet 2006;38: 594–597.

34 Messing J, Bharti AK, Karlowski WM, Gundlach H, Kim HR, et al: Sequence composition and genome organization of maize. Proc Natl Acad Sci USA 2004;101:14349–14354.

35 Piegu B, Guyot R, Picault N, Roulin A, Saniyal A, et al: Doubling genome size without polyploidization: dynamics of retrotransposition-driven genomic expansions in *Oryza australiensis*, a wild relative of rice. Genome Res 2006;16:1262–1269.

36 Morgante M: Plant genome organisation and diversity: the year of the junk! Curr Opin Biotechnol 2006;17:168–173.

37 Almeida J, Carpenter R, Robbins TP, Martin C, Coen ES: Genetic interactions underlying flower color patterns in *Antirrhinum majus*. Genes Dev 1989;3:1758–1767.

38 Dooner HK, Robbins TP, Jorgensen RA: Genetic and developmental control of anthocyanin biosynthesis. Annu Rev Genet 1991;25:173–199.

39 Kuang H, Ochoa OE, Nevo E, Michelmore RW: The disease resistance gene *Dm3* is infrequent in natural populations of *Lactuca serriola* due to deletions and frequent gene conversions at the RGC2 locus. Plant J 2006;47:38–48.

40 Bennetzen JL: Transposable element contributions to plant gene and genome evolution. Plant Mol Biol 2000;42:251–269.

41 Cheng C, Daigen M, Hirochika H: Epigenetic regulation of the rice retrotransposon Tos17. Mol Genet Genomics 2006;276:378–390.

42 Jiang N, Bao Z, Zhang X, Eddy SR, Wessler SR: Pack-MULE transposable elements mediate gene evolution in plants. Nature 2004;431:569–573.

43 Lisch D: Pack-MULEs: theft on a massive scale. Bioessays 2005;27:353–355.

44 Lal SK, Giroux MJ, Brendel V, Vallejos CE, Hannah LC: The maize genome contains a helitron insertion. Plant Cell 2003;15:381–391.

45 Kapitonov VV, Jurka J: Rolling-circle transposons in eukaryotes. Proc Natl Acad Sci USA 2001;98:8714–8719.

46 Gupta S, Gallavotti A, Stryker GA, Schmidt RJ, Lal SK: A novel class of *Helitron*-related transposable elements in maize contain portions of multiple pseudogenes. Plant Mol Biol 2005;57: 115–127.

47 Long M, de Souza SJ, Gilbert W: Evolution of the intron-exon structure of eukaryotic genes. Curr Opin Genet Dev 1995;5:774–778.

48 Patthy L: Genome evolution and the evolution of exon-shuffling – a review. Gene 1999;238: 103–114.

49 Liu M, Grigoriev A: Protein domains correlate strongly with exons in multiple eukaryotic genomes – evidence of exon shuffling? Trends Genet 2004;20:399–403.
50 Kato A, Lamb JC, Birchler JA: Chromosome painting using repetitive DNA sequences as probes for somatic chromosome identification in maize. Proc Natl Acad Sci USA 2004;101:13554–13559.
51 Ananiev EV, Chamberlin MA, Klaiber J, Svitashev S: Microsatellite megatracts in the maize (*Zea mays* L.) genome. Genome 2005;48:1061–1069.
52 Fu H, Zheng Z, Dooner HK: Recombination rates between adjacent genic and retrotransposon regions in maize vary by 2 orders of magnitude. Proc Natl Acad Sci USA 2002;99:1082–1087.
53 Anderson LK, Lai A, Stack SM, Rizzon C, Gaut BS: Uneven distribution of expressed sequence tag loci on maize pachytene chromosomes. Genome Res 2006;16:115–122.

Antoni Rafalski
DuPont Crop Genetics Research, DuPont Experimental Station Building E353
Route 141 and Henry Clay Road
Wilmington, DE 198803 (USA)
Tel. +1 302 695 2634, Fax +1 302 695 2726, E-Mail j-antoni.rafalski@cgr.dupont.com

Volff J-N (ed): Plant Genomes.
Genome Dyn. Basel, Karger, 2008, vol 4, pp 131–142

The Rice Genome Structure as a Trail from the Past to Beyond

T. Sasaki

National Institute of Agrobiological Sciences, Kannondai 2-chome, Tsukuba, Ibaraki, Japan

Abstract

Rice is the first cultivated plant species to be completely sequenced. Cultivation, in general, is a major factor that contributes to selection pressure of a target species resulting in accelerated changes in genome structure from an original wild species. The genome sequence of cultivated rice can now be used as a standard for comparison with many other cultivated rice species, its wild relatives and other grass species. Several indices define the dynamic nature of genome structure and function. Genome dynamics in rice is described here based on transposon, retrotransposon and polyploidy. Comparative studies with other related grass species based on the principle of synteny are expected to generate invaluable information that will clarify the trail that led to the present cultivated rice as well as the trail that will lead to its further improvement.

Plants carry their own genomes as resource of inheritable materials that determine their characteristic appearance and behavior throughout the life cycle. The genome represents the whole hereditary information encoded in the DNA or the nucleotide sequences of the chromosomes in corresponding plant species. Under natural conditions, this genetic information is inherited from one generation to another in a process that involves selection of traits, which could easily adapt to the environment and are more likely to survive. The genome structure could have been altered by a neutral selection process followed by a natural mutation, which occurred at a constant frequency particularly in cases of miscopy by DNA polymerase or exposure to ionizing radiation. Among cultivated plants, changes in genome structure must have been artificially induced by mankind through repeated selection and breeding. This procedure relied on finding natural mutations and selection of the most useful mutation which gave rise to breeding methods that have been adopted since the beginning of agriculture more than 10,000 years ago. The continuous pressure of selection has been so strong that it is almost impossible to find the ancestral species for some of the currently domesticated plant species. In addition to the selection of favorable mutation in the nucleotide sequences, unnatural selection process must have contributed to the activation of transposable elements. In cultivated plants, many genes interrupted by insertion of transposons have been clarified from extensive genome sequence analysis. Polyploidy is also known in many plant species arising from spontaneous somatic chromosome duplication (autoploidization), or as a result of non-disjunction of homeologous chromosomes (alloploidization). Although domestication in itself is not necessarily the major cause of polyploidy, many cultivated plant species carry multiple sets of chromosomes. Several known polyploids such as wheat, banana and tobacco are more vigorous than their diploid progenitors and are therefore highly preferred in agriculture. In case of cultivated rice, detailed genome sequence analysis has identified its chromosomal structure as a nearly pure diploid. However, some of the tetraploid wild *Oryza* species give lesser amounts of seeds than the diploid *O. sativa* [1]. This is paradoxical considering the general rule of parallel relationship between polyploidy and plant size.

These suggest that a lot of phenomena associated with the evolution of cultivated rice still remain to be identified. Although the genome must correctly inherit information from generation to generation, it is also susceptible to changes, which occurred spontaneously due to known and unknown factors that induce phenotypic mutations. If such changes had no effect on the maintenance of life, they remain as mutation or polymorphism in a species. However, if unknown strong forces induced drastic changes to the structure of the genome such as segmental duplication of chromosomes or insertion/deletion of transposable elements, but without any effect on the ability to survive, a new species might be generated. In case of *O. sativa*, the selection pressure by breeders and farmers may have contributed a great deal in defining specific characteristics as a response to the environment. The elucidation of the genome sequence of a representative rice cultivar may provide a clue for elucidating genome dynamics in cultivated rice, in other *Oryza* species and other cereal crops in general.

Genome Structure of *O. sativa* ssp. *japonica* Cultivar Nipponbare

The International Rice Genome Sequencing Project (IRGSP) [2] chose the *japonica* rice cultivar Nipponbare as a common template for complete and accurate decoding of the *Oryza sativa* genome [3]. This cultivar was one of the parent cultivars of the F_2 population used for linkage analysis to construct a high-density molecular genetic map of rice [4]. It was also used as a resource for construction of cDNA libraries to generate a catalog of rice ESTs [5] and full-length cDNAs [6]. About 6,000 of these transcripts were used as genetic and/or PCR-based markers to reconstruct the rice genome using large-sized rice genomic DNA fragments ligated in yeast artificial chromosome (YAC), bacterial artificial chromosome (BAC), and P1-derived artificial chromosome (PAC) vectors [7, 8, http://rgp.dna.affrc.go.jp/cgi-bin/statusdb/irgsp-status.cgi?lang = en]. Additionally, the *japonica* cultivar Nipponbare can be regenerated easily from callus culture and transformed by polyethylene glycol or *Agrobacterium*-mediated methods [9, 10] making it advantageous to prove gene function after genome sequencing. The IRGSP was organized in 1998 to pursue the sequencing of the genome and eventually succeeded in completely decoding the rice genome sequence with 99.99% accuracy [2]. The major features of the rice genome based on the high-quality genome sequence are as follows:

(1) The genome size of Nipponbare is 389 Mb including the unsequenced regions or gaps in the physical map, which were measured by fiber-FISH and conventional FISH.

(2) The genome consists of a total of 37,544 non-transposable element-related genes, which were predicted *ab initio* by FGENESH after masking the repeat sequences.

(3) Twenty nine percent of the genes (10,837) are duplicated at least once in tandem within an adjacent 5-Mb region.

(4) Chloroplast and mitochondrial insertions contribute 0.20–0.24% and 0.18–0.19% of the nuclear genome, respectively, depending upon the stringency adopted to identification of each organellar genome sequence.

(5) Class I elements, which include long terminal-repeat (LTR) retrotransposons (Ty1/*copia*, Ty3/*gypsy* and *TRIM*) and non-LTR retrotransposons (LINEs and SINEs, or long- and short-interspersed nucleotide elements, respectively), occupy 13% of the rice genome.

(6) Class II elements, characterized by terminal inverted-repeats and including the *h*AT, CACTA, IS256/*Mutator*, IS5/*Tourist*, and IS*630*/Tc1/*mariner* superfamilies, occupy 19.4% of the rice genome.

(7) A total of 18,828 class I simple sequence repeats (SSR) were identified as di-, tri-, and tetra-nucleotide SSRs.

After the publication of the genome sequence in 2005, continuous efforts have been carried out to complete the physical map by reducing the gaps and identifying clones containing telomere associated sequence, CCCTAAA [11]. The physical map of chromosome 5 centromere was completed but its sequence was only partly elucidated because of many repeat sequences. The basic structure so far analyzed is composed of a small number of large clusters of centromeric

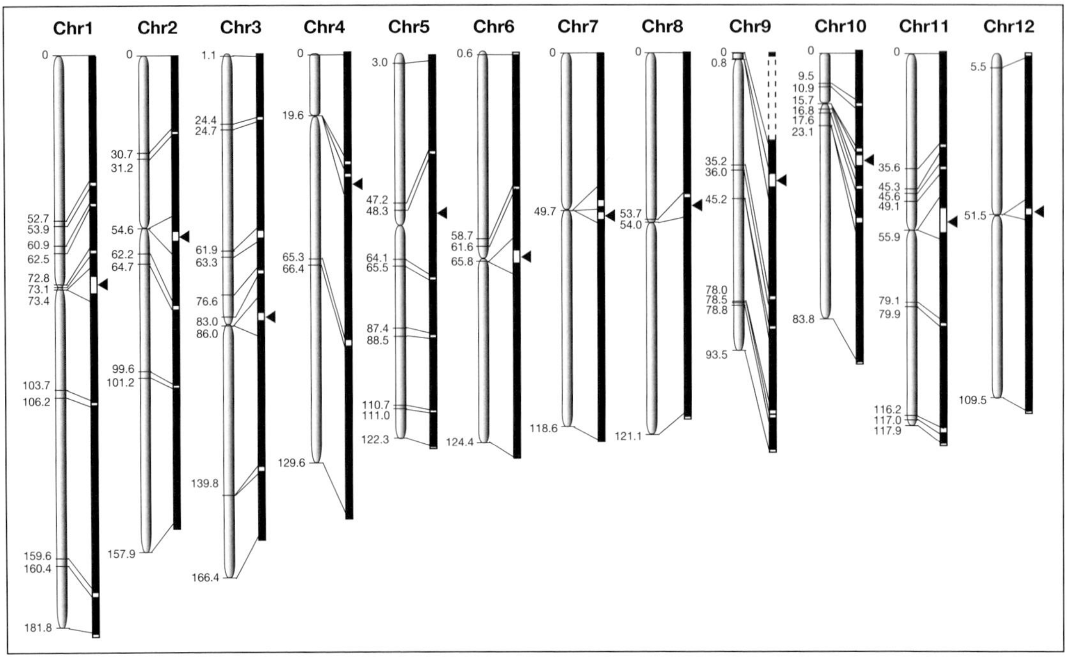

Fig. 1. Sequence-ready physical map constructed by the International Rice Genome Sequencing Project (IRGSP) with PAC and BAC clones. This map was revised in 2007. For each chromosome, the left and right bars represent the genetic map and physical map, respectively. The numbers at the left side of the genetic map indicate the genetic distance (cM) calculated by linkage analysis using an F2 population derived from crossing *japonica* rice cultivar Nipponbare and *indica* rice cultivar Kasalath. The arrow at the right side of each chromosomal physical map indicates the position of the centromere. The gaps on the physical map of each chromosome indicate uncovered regions, although the length does not correspond to the actual size of the gap.

satellite repeat DNA known as CentO, that is similar to that of chromosome 8. The latest physical map is shown in figure 1.

The detailed annotation of the completed rice genome sequence was carried out using rice full-length cDNAs as reference. This effort resulted in the accurate prediction of 29,550 expressed loci including the unmapped-mRNA clusters [12, http://rapdb.dna.affrc.go.jp/]. The details of the genome annotation can be accessed through the Rice Annotation Project Database (RAP-DB) with the gene locus consisting of 7 digits as in Os01gxxxxxxx, in which Os and 01g mean *O. sativa* and chromosome 1, respectively.

The existence of chromosome segmental duplication in the rice genome was clarified by a dot matrix plot of predicted gene products. In addition to the well-known duplication between the distal ends of chromosomes 11 and 12 [13], the existence of 10 duplicated segments between chromosomes 1 and 5, as well as between chromosomes 2 and 4 were also identified (fig. 2). These duplicated regions were not clarified by genetic analysis based on RFLP partly due to the absence of co-segregating polymorphic components of RFLP and partly due to the sparse density of homologous genes within the newly identified duplicated segment. The overall signature of duplication in the rice genome is less significant than that of *Arabidopsis thaliana* [14]. Polyploidy is generally observed among cultivated plants such as hexaploid wheat, tetraploid cotton and plausibly soybean. Even in weedy plants, segmental genome duplication is also observed as evidenced in the case of *Arabidopsis*. If such duplication generally occurs among plant species irrespective of whether they are cultivated or not, it will be necessary to identify the factor(s) that induced such events in the course of evolution. This may also provide evidence on the diploid nature of cultivated rice. If human being has selected polyploidy to get more yield, tetraploid *Oryza* should have been favored. However the only tetraploid type of *Oryza* is a weedy plant with large morphological phenotypes but small grains [15].

Retrotransposons in the Rice Genome

Retrotransposons are known to be stably inserted via RNA intermediate near or within genes thereby inducing mutations. Several types of retrotransposons characterized by long terminal repeats (LTRs) were found in the Nipponbare genome [16]. Among them, the copia-like retrotransposon *Tos17* was the most intensively analyzed [17] because of its moderate frequency of transposition thereby allowing accurate identification of inserted sequence. The number of authentic copies of *Tos17* varies among rice subspecies and cultivars, with the cultivar Nipponbare of the subspecies *japonica* containing 2 copies. The *Tos17* is activated under tissue culture condition and the frequency of transposition varies by cultivar. In the case of Nipponbare, after several months of callus culture, the regenerated plants carry approximately 10 copies of transposed *Tos17*. In case of another *japonica* cultivar Akitakomachi, transposed copy number after a similar duration of callus culture is more than that of Nipponbare (Miyao et al., unpublished data). Thus although transposition does not occur under natural conditions, *Tos17* can be effectively used for gene disruption to identify corresponding gene function. In addition, the flanking sequence of the inserted *Tos17* gives novel information on the preference of *Tos17* insertion in the rice genomic sequence. This statistical data should be

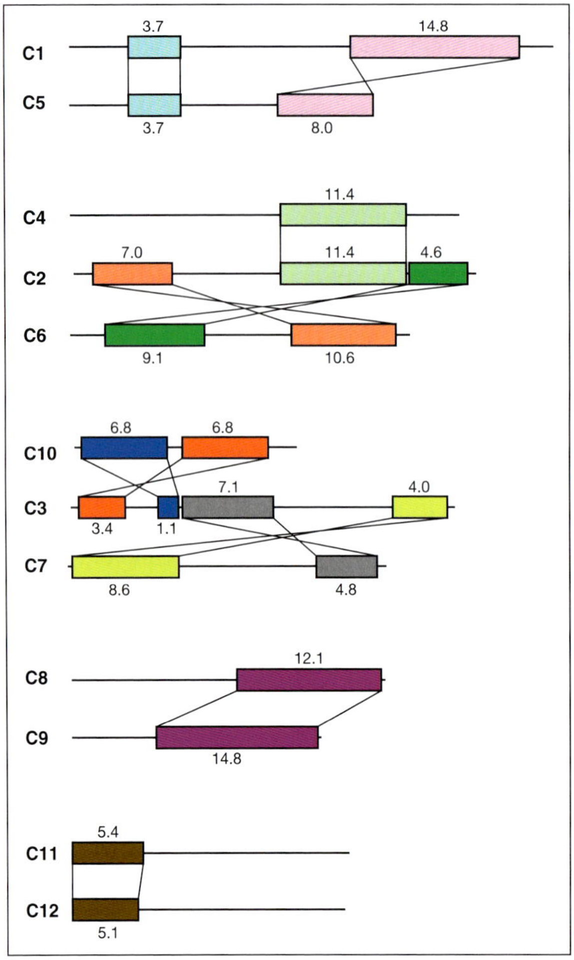

Fig. 2. Graphical representation of segmental duplication between chromosomes. Duplication was identified by a dot plot of predicted translated gene products [2]. Each horizontal line indicates chromosome and its numbering like C1 means chromosome 1. The left side of each chromosome corresponds to the top of physical map in figure 1. Each pair of colored boxes is the corresponding duplicated genomic segment. The lines connecting each colored box indicate the start and end of duplication. The crossing of lines means the reverse direction of the duplicated region in corresponding chromosomes. The number attached to each box shows the length (Mb) of the duplicated region.

useful to understand the interaction mechanism among *Tos17* sequence, targeted genome sequence, and enzymes involved in replication. As of July 2007, ca. 4000 genes have been identified as disrupted by at least one *Tos17* insertion among mutant plant lines. The distribution of disrupted genes along the 12 Nipponbare chromosomes and characteristics of disrupted sequences is available on http://tos.nias.affrc.go.jp/~miyao/pub/tos17/. The most frequently disrupted locus with 398 times insertion is Os02g0118800 which is annotated as NBS-LRR protein. The insertion target nucleotide sequences revealed a palindromic consensus sequence, ANGTT-(TSD, target site duplication sequence)-AACNT, flanking the 5-bp target site duplication [18]. Although the mechanism of *Tos17* insertion is not yet clearly understood, there were indications that methylation of histone H3K9 necessary for methylation of *Tos17* followed by transposition was involved [19].

DNA Transposons in the Rice Genome

Several types of transposons directly copied to other genomic positions are known in the rice genome [20]. Transposons are categorized into two groups, that is, non-autonomous transposons which lack enzymes required for transposition and autonomous transposons which encode these enzymes. Among non-autonomous transposons, MITE (miniature inverted-repeat transposable element) is characterized by its short terminal inverted repeat at both sequence ends of insertion. The Nipponbare genome contains ca. 90,000 copies of MITEs. As MITEs lack a system to facilitate their own transposition to other genomic positions, their transposition must require support by transposase activity. Recently identified *mPing*, one of the members of rice MITEs with a size of 430 bp lacks transposase, but its transposition was confirmed under cell culture and anther culture, respectively [21, 22]. Rice mutant with slender glume generated by gamma-ray irradiation was identified to be caused by an insertion of *mPing* into a gene, *rice ubiquitin-related modifier* 1 [23]. About 50 copies of *mPing* were found in the Nipponbare genome sequence. In addition, two types of DNA fragments were found which shared terminal inverted repeat sequence with *mPing*. They were named *Ping* (5,341 bp) and *Pong* (5,166 bp), respectively and both contained two ORFs [21]. One of the ORFs has the DDE motif commonly identified in transposase and is thought to be responsible for the transposition of *mPing*, by both of *Ping* and *Pong* [21, 22]. The copy number of *mPing* varies among rice cultivars. In general, about 50 copies are recognized, and in the most abundant case, about 1,100 copies are amplified. If this amplification can be controlled, *mPing* could also be used as an effective tool for functional analysis by gene disruption.

A 607-bp non-autonomous transposon, *nDart* was first identified in the *albino* rice mutant inserted into the gene encoding Mg-protoporphyrin IX methyltransferase [24]. Thereafter, this transposon was also characterized in a spontaneous mutable *virescent* mutant (*pyl-v*) inserted in a putative chloroplast protease, and in the dwarf mutant *thumbelina-mutable* (*thl-m*) inserted in a gene with unknown function [25]. At least 13 copies of *nDart* were identified in the Nipponbare genome sequence. In addition, at least 56 copies of a much larger *nDart* related sequence (average size of 3,584 bp) which contained one putative *Dart* ORF were also identified. The predicted gene product of this ORF shares significant similarity with the transposases of the *hAT* superfamily and contains a putative *hAT* dimerization motif. These ORFs in Nipponbare are expected to work as transposase, but Nipponbare does not excise *nDart* from the above mutant lines. The same situation was observed in the *indica* cultivar Kasalath. One of the parental lines of *pyl-v*, a *japonica* cultivar H-126 which is also one of the parental lines of *pyl-v* mutant could excise the inserted *nDart* to give a stable pale-yellow leaf mutant. This mutant was thought to carry an autonomous transposon, *aDart* [25]. Unlike *Tos17* activation in rice, *Dart* might be useful as a tool for tagging genes by avoiding somaclonal variation accompanied with cell culture.

Polyploidy in *Oryza* Species

Polyploidy is widely observed in plants especially among cultivated crops as a result of continuous hybridization and modifications by humans. Most polyploid plants are highly favored in agriculture because of larger size. Autopolyploidy could be generated using colchicine, but this type of polyploidy is practically less useful than allopolyploidy which is spontaneously generated, or by artificial crossing of closely related genomes. Polyploid plants with odd numbers of chromosomes such as banana do not produce seeds and are propagated vegetatively. On the other hand, polyploids with even number of chromosomes such as wheat can produce seeds and can be easily crossed and propagated.

In the genus *Oryza*, 9 types of genomes, namely, AA, BB, CC, BBCC, CCDD, EE, FF, GG, and HHJJ are known (http://www.shigen.nig.ac.jp/rice/oryzabase/). Widely cultivated rice species consist mainly of the AA genome and none of the other genomes is practically used. However, introgression between AA genome by crossing is possible and embryo rescue method is applicable between cultivated AA genome and other genomes to transfer alien DNA segments [1]. This procedure has been successfully carried out to take advantage of the existence of biotic or abiotic stress tolerance genes in wild species of *Oryza*. For example, introgression of *O. latifolia* with CCDD genome was performed to transfer resistance genes to bacterial blight, brown planthopper and the whitebacked planthopper [26].

The AA genome species of *Oryza sativa* must have been chosen as the most suitable rice crop for cultivation thousands of years ago. As a result only cultivated subspecies of *O. sativa* remain today. Otherwise, wild *O. sativa* changed its original genome by continuous breeding pressure and currently it can survive only under cultivation by farmers. This is generally observed in many domesticated species. The currently available ancestral wild relatives of *O. sativa* are thought to be *O. nivara* and *O. rufipogon* [27]. Both of them are diploid and widely grow from the tropical and subtropical areas in Asia and Oceania. Many intra-species variations also exist to enable 'evolution' by breeding pressure. The non-shattering habit which is one of the definitive characteristics of domesticated rice plants was partly identified as a polymorphism of amino acid sequence of the protein involved in abscission layer formation [28]. Also another locus controlling shattering between *japonica* and *indica* subspecies was identified as SNP in the 5′ regulatory region of a BEL1-type homeobox gene [29]. This SNP gives difference in the formation of abscission layer. Another factor of domestication, high yield, is also a complex character involving many genetically defined components. The number of publications on rice yield is so far the highest rank among the traits for rice. The accumulated data shows that almost all of the genomic regions are involved in genetics of yield. This indicates that many genes in rice contribute to the expression of the character of 'yield'. Among such genes, a recently identified gene involved in grain number may provide a clue to anatomize the network of cytokinin, one of the plant hormones involved in cell differentiation at inflorescence meristem [30]. The detailed molecular genetic analyses of rice heading date, which is an important characteristic to increase yield have also been successfully carried out [31, 32]. These information and genes themselves now make it possible to design rice plants with highly desirable characteristics by both conventional and non-conventional methods of genetic modification.

The tetraploid *Oryza* species known so far such as BBCC, CCDD, HHKK, and HHJJ genomes are hybrids between the phylogenetically closely related species [33]. Interestingly, no hybrids between AA and BB genomes are known. The *Oryza* species with AA genome consists of currently cultivated rice plants. Since hybrid species must have evolved after the establishment of each corresponding species, some molecular mechanisms may have been involved in prohibiting or promoting hybrid formation and survival between different types of genomes. Among cereal crops, hybrid formation is well known in wheat, *Triticum aestivum*. Its genome is composed of AABBDD and related wild species with AA, BB, DD and AABB genomes are known. The common wheat *T. aestivum* was believed to be generated by a spontaneous hybridization between AA and BB genome followed by a spontaneous hybridization of DD with AABB genome species. In the wheat genome, there exists a very important key gene, *Ph1* that prevents chromosome pairing among related genome components during meiosis. Recent molecular analysis of *Ph1* could narrow down its location within a 2.5 Mb region of chromosome 5B [34]. This region shows genomic co-linearity with the corresponding region of rice chromosomes 8 and 9 because of internal duplication of rice genome. So far no wheat orthologous *Ph1* has ever been reported in tetraploid *Oryza* species. However, without such mechanism as *Ph1*, correct pairing of original chromosomes during meiosis is expected. In diploid *O. sativa*, key genes which are required for homologous chromosome pairing in meiosis were identified and named *PAIR1* and *PAIR2*, respectively [35, 36]. If diploid *Oryza* with BB genome is modified regarding genes of preferable yield, and this modified BB genome rice is hybridized with an elite variety AA genome using the above mentioned information, tetraploid rice with high-yield and much tolerance to stress could be generated.

Conclusion

It is well known that cultivated plants can be easily subjected to various breeding programs to produce new varieties within a short period. If only sequence variation which spontaneously occurred is reflected in breeding, it is probably impossible to sustain a stable food supply for mankind. Variation of phenotypes is derived from variation of nucleotide sequences corresponding to the phenotype. In general, evolution is propelled by the occurrence of random mutation in nucleotide sequence and subsequent adaptation to the environment. This is true particularly among naturally living organisms. However, many cereal crops, vegetables and ornamental plants are constantly subjected to intentional and continuous breeding selection pressure resulting in increased chance of generating variation in targeted traits. Highly desirable traits were selected and subsequent crossing with other cultivars could generate the desired variation. Under normal circumstances, the genome of each organism is both conservative, facilitating stable inheritance of genetic information, and innovative, allowing the organism to adjust to the environment and compete with others. Current breeding strategies rely on the plasticity of genome. In addition to genome dynamism, two other phenomena, namely, heterosis and gene silencing are important for agriculture. Heterosis, which is known as the increased strength of different characteristics in hybrids, is widely used in modern agriculture and many crops are currently propagated as F_1 seeds. However, the molecular nature of heterosis is still unclear, although several ideas have been proposed [37]. On the other hand, the mechanism of gene silencing is understood in detail and novel finding on RNA function was identified by this phenomenon and its biological importance is widely recognized in all of the higher organisms including plants [38]. The mechanism of gene silencing must be useful to suppress virus infection to plants. Currently, we can manipulate rice genome not only by conventional genetic method but also by vector-aided transformation method. For these manipulations, plasticity of the genome is prerequisite to effectively introduce alien genes. Rice has been grown for a long time partly because of easiness to induce mutation preferable to cultivation and business. We now have a standard tool for developing the future potential of rice genomics-based research for meeting the challenges of cereal crop production.

References

1 Brar DS, Khush GS: Alien introgression in rice. Plant Mol Biol 1997;35:35–47.

2 International Rice Genome Sequencing Project: The map-based sequence of the rice genome. Nature 2005;436:793–800.

3 Sasaki T, Burr B: International Rice Genome Sequencing Project: the effort to completely sequence the rice genome. Curr Opin Plant Biol 2000;3:138–141.

4 Harushima Y, Yano M, Shomura A, Sato M, Shimano T, et al: A high-density rice genetic linkage map with 2275 markers using a single F2 population. Genetics 1998;148:479–494.

5 Sasaki T, Song J, Koga-Ban Y, Matsui E, Fang F, et al: Toward cataloguing all rice genes: Large scale sequencing of randomly chosen rice cDNAs from a callus cDNA library. Plant J 1994;6:615–624.

6 The Rice Full-Length cDNA Consortium: Collection, mapping, and annotation of over 28,000 cDNA clones from *japonica* rice. Science 2003;301:376–379.

7 Baba T, Katagiri S, Tanoue H, Tanaka R, Chiden Y, et al: Construction and characterization of rice genomic libraries: PAC library of *japonica* variety, Nipponbare and BAC library of *indica* variety, Kasalath. Bull NIAR 2000;14:4–49.

8 Wu J, Maehara T, Shimokawa T, Yamamoto S, Harada C, et al: A comprehensive rice transcript map containing 6591 expressed sequence tag sites. Plant Cell 2002;14:525–535.

9 Hayashimoto A, Li Z, Murai N: A polyethylene glycol-mediated protoplast transformation system for production of fertile transgenic rice plants. Plant Physiol 1990;93:857–863.

10 Raineri DM, Bottino P, Gordon MP, Nester EW: *Agrobacterium*-mediated transformation of rice (*Oryza sativa L.*). Nat Biotech 1990;8:33–38.

11 Mizuno H, Wu J, Kanamori H, Fujisawa M, Namiki N, et al: Sequencing and characterization of telomere and subtelomere regions on rice chromosomes 1S, 2S, 2L, 6L, 7S, 7L, and 8S. Plant J 2006;46:206–217.

12 The Rice Annotation Project: Curated genome annotation of *Oryza sativa* ssp. *japonica* and comparative genome analysis with *Arabidopsis thaliana*. Genome Res 2007;17:175–183.

13 Wu J, Kurata N, Tanoue H, Shimokawa T, Umehara Y, Yano M, Sasaki T: Physical mapping of duplicated regions of two chromosome ends in rice. Genetics 1998;150:1595–1603.

14 The Arabidopsis Genome Initiative: Analysis of the genome sequence of the flowering plant *Arabidopsis thaliana*. Nature 2000;408:796–815.

15 Vaughan DA: 'The Wild Relatives of Rice'. International Rice Research Institute, Los Banos, Phillipines, 1994.

16 Hirochika H, Fukuchi A, Kikuchi A: Retrotransposon families in rice. Mol Gen Genet 1992;233:209–216.

17 Hirochika H, Sugimoto K, Otsuki Y, Tsugawa H, Kanda M: Retrotransposons of rice involved in mutations induced by tissue culture. Proc Natl Acad Sci USA 1996;93:7783–7788.

18 Miyao A, Tanaka K, Murata K, Sawaki H, Takeda S, et al: Target site specificity of the *Tos17* retrotransposon shows a preference for insertion within genes and against insertion in retrotransposon-rich regions of the genome. Plant Cell 2003;15:1771–1780.

19 Ding Y, Wang X, Su L, Zhai JX, Cao SY, et al: SDG714, a histone H3K9 methyltransferase, is involved in *Tos17* DNA methylation and transposition in rice. Plant Cell 2007;19:9–22.

20 Turcotte K, Srinivasan S, Bureau T: Survey of transposable elements from rice genome sequence. Plant J 2001;25:169–179.
21 Jiang N, Bao Z, Zhang X, Hirochika H, Eddy SR, McCouch SR, Wessler SR: An active DNA transposon family in rice. Nature 2003;421:163–167.
22 Kikuchi K, Terauchi K, Wada M, Hirano H: The plant MITE *mPing* is mobilized in anther culture. Nature 2003;421:167–170.
23 Nakazaki T, Okumoto Y, Horibata A, Yamahira S, Teraishi M, et al: Mobilization of a transposon in the rice genome. Nature 2003;421:170–172.
24 Fujino K, Sekiguchi H, Kiguchi T: Identification of an active transposon in intact rice plants. Mol Gen Genomics 2005;273:150–157.
25 Tsugane K, Maekawa M, Takagi K, Takahara H, Qian Q, Eun CH, Iida S: An active DNA transposon nDart causing leaf variegation and mutable dwarfism and its related element in rice. Plant J 2006;45:46–57.
26 Multani DS, Khush GS, de los Reyes BG, Brar DS: Alien genes introgression and development of monosomic alien addition lines from *Oryza latifolia* Desv. to rice, *Oryza sativa* L. Theor Appl Genet 2003;107:395–405.
27 Khush GS: Origin, dispersal, cultivation and variation of rice. Plant Mol Biol 1997;35:25–34.
28 Li C, Zhou A, Sang T: Rice domestication by reducing shattering. Science 2006;311:1936–1939.
29 Konishi S, Izawa T, Lin SY, Ebana K, Fukuta Y, Sasaki T, Yano M: An SNP caused loss of seed shattering during rice domestication. Science 2006;312:1392–1396.
30 Ashikari M, Sakakibara H, Lin S, Yamamoto T, Takashi T, et al: Cytokinin oxidase regulates rice grain production. Science 2005;309:741–745.
31 Yano M, Katayose Y, Ashikari M, Yamanouchi U, Monna L, et al: *Hd1*, a major photoperiod sensitivity quantitative trait locus in rice, is closely related to the *Arabidopsis* flowering time gene *CONSTANS*. Plant Cell 2000;12:2473–2483.
32 Kojima S, Takahashi Y, Kobayashi Y, Monna L, Sasaki T, Araki T, Yano M: *Hd3a*, a rice ortholog of the *Arabidopsis FT* gene, promotes transition to flowering downstream of *Hd1* under short-day conditions. Plant Cell Physiol 2002;43:1096–1105.
33 Ge S, Sang T, Lu BR, Hong DY: Phylogeny of rice genomes with emphasis on origins of allotetraploid species. Proc Natl Acad Sci USA 1999;96:14400–14405.
34 Griffith S, Sharp R, Foote TN, Bertin I, Wanous M, et al: Molecular characterization of *Ph1* as a major chromosome pairing locus in polyploid wheat. Nature 2006;439:749–752.
35 Nonomura K, Nakano M, Fukuda T, Eiguchi M, Miyao A, Hirochika H, Kurata N: The novel gene *HOMOLOGOUS PAIRING ABERRATION IN RICE MEIOSIS 1* of rice encodes a putative coiled-coil protein required for homologous chromosome pairing in meiosis. Plant Cell 2004;16:1008–1020.
36 Nonomura KI, Nakano M, Murata K, Miyoshi K, Eiguchi M, et al: An insertional mutation in the rice *PAIR2* gene, the ortholog of *Arabidopsis ASY1*, results in a defect in homologous chromosome pairing during meiosis. Mol Gen Genomics 2004;271:121–129.
37 Lipman ZB, Zamir D: Heterosis: revisiting the magic. Trends Genet 2007;23:60–66.
38 Baulcombe D: RNA silencing in plants. Nature 2004;431:356–363.

Takuji Sasaki
National Institute of Agrobiological Sciences
1–2, Kannondai 2-chome
Tsukuba, Ibaraki 305–8602 (Japan)
Tel. +81 29 838 7097, Fax +81 29 838 7075, E-Mail tsasaki@nias.affrc.go.jp

Author Index

Subject Index